D1417842

## STUDIES IN CHEMICAL PHYSICS

**General Editor**

A.D. Buckingham F.R.S., Professor of Chemistry, University of Cambridge

*Series Foreword*

The field of science known as 'Chemical Physics' has greatly expanded in recent years. It is an essential part of both physics and chemistry and now impinges on biology, crystallography, the science of materials and even on astronomy. The aim of this series is to present short, authoritative and readable books on different topics in chemical physics at a level that is appreciated by the non-specialist and yet is of prime interest to the expert–in fact, the type of book that we all welcome and enjoy.

I was grateful to be given the opportunity to help plan this series, and warmly thank the authors and publishers whose efforts have brought it into being.

A.D. Buckingham,
University Chemical Laboratory,
Cambridge, U.K.

*Other titles in the series*

*Advanced Molecular Quantum Mechanics,* R.E. Moss
*Chemical Applications of Molecular Beam Scattering,* M.A.D. Fluendy and K.P. Lawley
*Electronic Transitions and the High Pressure Chemistry and Physics of Solids,* H.G. Drickamer and C.W. Frank
*Aqueous Dielectrics,* J.B. Hasted
*Statistical Thermodynamics,* B.J.M. McClelland

# Principles of Mössbauer Spectroscopy

T.C. GIBB

*Department of Inorganic and Structural Chemistry,*
*University of Leeds*

CHAPMAN AND HALL · LONDON

A HALSTED PRESS BOOK
JOHN WILEY & SONS INC., NEW YORK

*Distributed in the U.S.A. by Halsted Press*
*a Division of John Wiley & Sons, Inc., New York.*

**Library of Congress Cataloging in Publication Data**

Gibb, Terence Charles.
Principles of Mössbauer spectroscopy.

"A Halsted Press book."
Bibliography: p.
Includes index.
1. Mössbauer spectroscopy. I. Title.
QC491.G52 537.5′352 75–25878
ISBN 0 470-29743-3

*First published 1976*
*by Chapman and Hall Ltd,*
*11 New Fetter Lane, London EC4P 4EE*



*Typeset by Mid-County Press*
*and printed in Great Britain by*
*Fletcher & Son Ltd, Norwich*

*ISBN 0 412 13960 X*



*Distributed in the U.S.A. by Halsted Press*
*a Division of John Wiley & Sons, Inc., New York*

# Preface

The emergence of Mössbauer spectroscopy as an important experimental technique for the study of solids has resulted in a wide range of applications in chemistry, physics, metallurgy and biophysics. This book is intended to summarize the elementary principles of the technique at a level appropriate to the advanced student or experienced chemist requiring a moderately comprehensive but basically non-mathematical introduction. Thus the major part of the book is concerned with the practical applications of Mössbauer spectroscopy, using carefully selected examples to illustrate the concepts. The references cited and the bibliography are intended to provide a bridge to the main literature for those who subsequently require a deeper knowledge.

The text is complementary to the longer research monograph, 'Mössbauer Spectroscopy', which was written a few years ago in co-authorship with Professor N.N. Greenwood, and to whom I am deeply indebted for reading the preliminary draft of the present volume. I also wish to thank my many colleagues over the past ten years, and in particular Dr. R. Greatrex, for the many stimulating discussions which we have had together. However my greatest debt is to my wife, who not only had to tolerate my eccentricities during the gestation period, but being a chemist herself was also able to provide much useful criticism of the penultimate draft.

Leeds, May 1975 T.C. Gibb

# Contents

CHAPTER ONE

# *The Mössbauer Effect*

The study of recoilless nuclear resonant absorption or fluorescence is more commonly known as Mössbauer spectroscopy. From its first origins in 1957, it has grown rapidly to become one of the most important research methods in solid-state physics and chemistry.

Resonant nuclear processes had been looked for without success for nearly thirty years before R.L. Mössbauer made his first accidental observation of recoilless resonant absorption in $^{191}$Ir in 1957 [1]. He not only produced a theoretical explanation of the effect which now bears his name, but also devised an elegant experiment which today remains almost unmodified as the primary technique of Mössbauer spectroscopy.

The Mössbauer effect is of fundamental importance in that it provides a means of measuring some of the comparatively weak interactions between the nucleus and the surrounding electrons. Although the effect is only observed in the solid state, it is precisely in this area that some of the most exciting advances in chemistry and physics are being made. Because it is specific to a particular atomic nucleus, such problems as the electronic structure of impurity atoms in alloys, the after-effects of nuclear decay, and the nature of the active-centres in iron-bearing proteins are just a few of the diverse and many applications.

## 1.1 Resonant absorption and fluorescence

Before delving into the details of the subject, it is worthwhile considering the historical perspective of what has come to be considered as a discovery of prime importance.

Atomic resonant fluorescence was predicted and discovered just after the turn of the century, and within a few years the underlying theory had been developed. From a simplified viewpoint, an atom in an excited electronic state can decay to its ground state by the emission of a photon to carry off the excess energy. This photon can then be absorbed by a second atom of the same kind by electronic excitation. Subsequent de-excitation re-emits the photon, but not necessarily in the initial direction so that scattering or resonant fluorescence occurs. Thus if the monochromatic yellow light from a sodium lamp is collimated and passed through a glass vessel containing sodium vapour, one would expect to see a yellow glow as the incident beam is scattered by resonant fluorescence.

A close parallel can be drawn between atomic and nuclear resonant absorption. The primary decay of the majority of radioactive nuclides produces a daughter nucleus which is in a highly excited state. The latter then de-excites by emitting a series of $\gamma$-ray photons until by one or more routes, depending on the complexity of the $\gamma$-cascade, it reaches a stable ground state. This is clearly analogous to electronic de-excitation, the main difference being in the much higher energies involved in nuclear transitions. It was recognized in the 1920's that it should be possible to use the $\gamma$-ray emitted during a transition to a nuclear ground-state to excite a second stable nucleus of the same isotope, thus giving rise to nuclear resonant absorption and fluorescence.

The first experiments to detect these resonant processes by Kuhn in 1929 [2] were a failure, although it was already recognized that the nuclear recoil and Doppler broadening effects (to be discussed shortly) were probably responsible. Continuing attempts to observe nuclear resonant absorption [3] were inspired by the realization that the emitted $\gamma$-rays should be an unusually good source of monochromatic radiation. This can easily be shown from the Heisenberg uncertainty principle. The ground state of the nucleus has an infinite lifetime and therefore there is no uncertainty in its energy. The uncertainty in the lifetime of the excited state is given by its mean life $\tau$, and the uncertainty in its energy is given by the width of the statistical energy distribution at half-height $\Gamma$. They are related by

$$\Gamma\tau \geqslant \hbar \tag{1.1}$$

where $h$ $(=2\pi\hbar)$ is Planck's constant. $\tau$ is related to the more familiar half-life of the state by $\tau = \ln 2 \times t_{1/2}$. If $\Gamma$ is given in electron-volts (1 eV = $1.60219 \times 10^{-19}$ J and is equivalent to 96.49 kJ $\text{mol}^{-1}$) and $t_{1/2}$ in seconds, then

$$\Gamma = 4.562 \times 10^{-16}/t_{1/2} \tag{1.2}$$

For a typical nuclear excited-state half-life of $t_{1/2} = 10^{-7}$ s, $\Gamma = 4.562 \times 10^{-9}$ eV. If the energy of the excited state is 45.62 keV, the emitted $\gamma$-ray will have an intrinsic resolution of 1 part in $10^{13}$. It should be borne in mind that the maximum resolution obtained in atomic line spectra is only about 1 in $10^8$.

The nucleus is normally considered to be independent of the chemical state of the atom because of the great disparity between nuclear and chemical energies. However, the $\gamma$-ray energy is so sharply defined that $\Gamma$ is smaller in magnitude than some of the interactions of the nucleus with its chemical environment. The typical magnitudes of some of the energies involved are indicated in Table 1.1.

Unfortunately there are several mechanisms which can degrade the energy of the emitted $\gamma$-ray, particularly the effects of the nuclear recoil and thermal energy [3]. Consider an isolated nucleus of mass $M$ with an excited state level at an energy $E$ and moving with a velocity $V$ along the direction in which the $\gamma$-ray is to be emitted (the components of motion perpendicular to this direction remain unaffected by the decay and may be ignored). The energy above the ground state *at rest* is $(E + \frac{1}{2}MV^2)$. When a $\gamma$-ray of energy $E_\gamma$ is emitted, the nucleus recoils and has a new velocity $V + v$ (which is a vector sum in that $V$ and $v$ may be opposed) and a total energy of $\frac{1}{2}M(V + v)^2$. By conservation of energy

$$E + \tfrac{1}{2}MV^2 = E_\gamma + \tfrac{1}{2}M(V + v)^2 \tag{1.3}$$

Table 1.1 Typical energies of nuclear and chemical interactions

| | | |
|---|---|---|
| Mössbauer $\gamma$-ray energies ($E_\gamma$) | $10^6$–$10^7$ | kJ mol$^{-1}$ |
| Chemical bonds and lattice energies | $10^2$–$10^3$ | kJ mol$^{-1}$ |
| Electronic transitions | 50–500 | kJ mol$^{-1}$ |
| Molecular vibrations | 5–50 | kJ mol$^{-1}$ |
| Lattice vibrations | 0.5–5 | kJ mol$^{-1}$ |
| Nuclear recoil and Doppler energies ($E_R$, $E_D$) | $10^{-2}$–1 | kJ mol$^{-1}$ |
| Nuclear quadrupole coupling constants | $<10^{-3}$ | kJ mol$^{-1}$ |
| Nuclear Zeeman splittings | $<10^{-3}$ | kJ mol$^{-1}$ |
| Heisenberg linewidths ($\Gamma$) | $10^{-7}$–$10^{-4}$ | kJ mol$^{-1}$ |

1 eV = $1.60219 \times 10^{-19}$ J and is equivalent to 96.49 kJ mol$^{-1}$

so that the actual energy of the photon emitted is given by

$$E_\gamma = E - \tfrac{1}{2}Mv^2 - MvV$$

$$= E - E_R - E_D \tag{1.4}$$

The $\gamma$-ray is thus deficient in energy by a recoil kinetic energy ($E_R = \frac{1}{2}Mv^2$) which is independent of the initial velocity $V$, and by a thermal or Doppler energy ($E_D = MvV$) which depends on $V$ and can therefore be positive or negative.

Momentum must also be conserved in the emission process. The momentum of the photon is $E_\gamma/c$ where $c$ is the velocity of light, so that

$$MV = M(V + v) + E_\gamma/c \tag{1.5}$$

and the recoil momentum is $Mv = -E_\gamma/c$. Hence the recoil energy is given by

$$E_R = \frac{E_\gamma^2}{2Mc^2} \tag{1.6}$$

and depends on the mass of the nucleus and the energy of the $\gamma$-ray. However, the Doppler energy $E_D$ is dependent on the thermal motion of the nucleus, and will therefore have a distribution of values which is temperature dependent. A mean value, $E_D$, can be defined which is related to the mean kinetic energy per translational degree of freedom, $\overline{E_k} \simeq \frac{1}{2}kT$, by

$$\overline{E_D} = 2\sqrt{(\overline{E_k}E_R)} = E_\gamma \sqrt{\frac{2\overline{E_k}}{Mc^2}} \tag{1.7}$$

where $k$ is Boltzmann's constant and $T$ is the absolute temperature.

As a result, the statistical distribution in energy of the emitted $\gamma$-rays is displaced from the true excited-state energy by $-E_R$ and broadened by $E_D$ into a Gaussian distribution of width $2\overline{E_D}$. The distribution for absorption has the same shape but is displaced by $+E_R$. This is illustrated schematically in Fig.1.1, and the order of magnitude of $E_R$ and $E_D$ is indicated in Table 1.1. Nuclear resonant absorption (which in practice can only be observed following an emission) will only have a significant probability if the emission and absorption energy distributions overlap strongly. The recoil and thermal broadening clearly prevent this. It is possible to compensate partially for $E_R$ by moving the emitter towards the observer at a very large (supersonic) velocity, or to increase the Doppler broadening by raising the temperature. Both result in only a marginal increase in the resonant overlap. Furthermore, the intrinsic resolution in the energy of the photon has been degraded to about 1 in $10^7$.

It should be noted that these equations are not peculiar to nuclear processes and apply equally well to atomic absorption. However, in the latter case the low energy of the transition results in the recoil energy being significantly less than the thermal broadening. Consequently the overlap for absorption is large and the effects of recoil may be ignored.

## 1.2 The Mössbauer effect

The Mössbauer effect is unique in that it provides a means of *eliminating* the destructive effects of the recoil and thermal energies. The key to the problem lies in the behaviour of the recoiling nucleus when it is no longer isolated (as is implicit in the preceeding discussion), but instead is fixed in a crystal lattice. As can be seen from Table 1.1, the recoil energy is much less than the chemical binding energy, but is similar in magnitude to the lattice-vibration phonon energies. If the recoil energy is transferred directly to

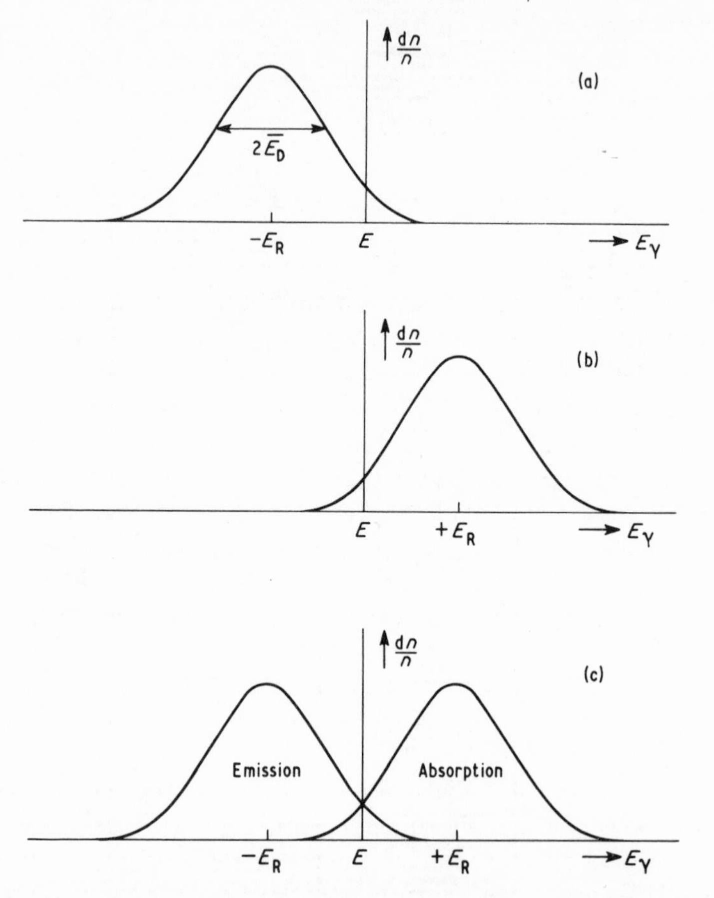


Fig. 1.1 The statistical distribution in the $\gamma$-ray energy for (a) emission, (b) absorption, and (c) the resonant overlap for successive emission and absorption.

vibrational energy, then the $\gamma$-ray energy is still degraded. However, the phonon energies are quantized, and the recoil energy can only be transferred to the lattice if it corresponds closely to an allowed quantum jump.

The simplest mathematical treatment is one in which the vibrational characteristics correspond to an Einstein solid with one vibrational frequency, $\omega$. Transfer of energy to the lattice can only take place in integral multiples of $\hbar\omega$ ($0, \pm\hbar\omega, \pm 2\hbar\omega, \ldots$). If the recoil energy $E_R$ is less than $\hbar\omega$, then either zero or one unit ($\hbar\omega$) of vibrational energy may be transferred to the lattice. It has been shown by Lipkin [4] that when many emission processes are considered, the *average* energy transferred per event must exactly equal $E_R$. If a fraction, $f$, of emission events result in no transfer of energy to the lattice (zero-phonon transitions), and a fraction $(1 - f)$ transfer one phonon of energy $\hbar\omega$, then

$$E_R = (1 - f)\hbar\omega$$

or

$$f = 1 - E_R/(\hbar\omega) \tag{1.8}$$

In a zero-phonon emission, the whole crystal rather than a single nucleus recoils. Equations (1.6) and (1.7) for $E_R$ and $\overline{E_D}$ contain the reciprocal mass, $1/M$. If the mass $M$ is increased to that of a crystallite containing perhaps $10^{15}$ atoms, then both the recoil energy and the Doppler broadening become very small and much less than $\Gamma$. Hence zero-phonon transitions are referred to as *recoilless*. In any real solid there will be a wide range of lattice frequencies, but fortunately it is very difficult to excite the low frequencies and there is still a significant fraction of nuclear events in which the $\gamma$-ray energy is not degraded.

To summarize, the Mössbauer effect is the resonant emission or absorption of $\gamma$-rays *in a solid matrix* without degradation by recoil or thermal broadening, and it gives an energy distribution dictated by the Heisenberg uncertainty principle.

The parameter $f$ is known as the recoilless or recoil-free fraction, and to increase the relative strength of the recoilless resonant process it is important that $f$ be as large as possible. On a more quantitative theory [5, 6] it is possible to relate $f$ to the vibrational properties of the crystal lattice by

$$f = \exp\left(-\frac{E_\gamma^2 \langle x^2 \rangle}{(\hbar c)^2}\right) \tag{1.9}$$

where $\langle x^2 \rangle$ is the mean-square vibrational amplitude of the nucleus in the direction of the $\gamma$-ray. From the form of the exponent, $f$ will only be large for a tightly bound atom with a small mean-square displacement, and for a small value of the $\gamma$-ray energy, $E_\gamma$. For the latter reason the highest energy for which a Mössbauer resonance is known is 187 keV in $^{190}Os$.

The precise form of $\langle x^2 \rangle$ depends on the vibrational properties of the lattice. In any real solid these are usually complex and adequate models are not available. However, it is instructive to consider the behaviour predicted [6] by the Debye model, which embodies a continuum of vibrational frequencies in the harmonic oscillator approximation following the distribution $N(\omega)$ = constant $\times\ \omega^2$, up to a maximum value of $\omega_D$. A characteristic temperature $\theta_D$ is defined by $\hbar\omega_D = k\theta_D$ and is known as the Debye temperature. The final expression obtained is

$$f = \exp\left\{-\frac{6E_R}{k\theta_D}\left[\frac{1}{4} + \left(\frac{T}{\theta_D}\right)^2 \int_0^{\theta_D/T} \frac{x\,dx}{e^x - 1}\right]\right\} \tag{1.10}$$

From the form of the equation it can be seen that $f$ is large when $\theta_D$ is large (a strong lattice) and when the temperature $T$ is small. This decrease in $f$ with rise in temperature follows in general from equation (1.9), in that thermal energy will increase the amplitude of vibration of the nucleus, but the precise form of the temperature dependence can only be predicted using a model for the lattice vibrations. This problem is discussed in more detail in Chapter 6.

It should be noted that although the recoilless fraction can be increased by lowering the temperature there is a limiting value at absolute zero which is still less than unity. For instance, from equation (1.10) when $T = 0$

$$f_{T=0} = \exp\left(-\frac{3E_R}{2k\theta_D}\right) \tag{1.11}$$

which still depends on $E_R$ and $\theta_D$.

In summary, recoilless emission or absorption is optimized for a low-energy $\gamma$-ray with the nucleus strongly bound in a crystal lattice at low temperature.

## 1.3 The Mössbauer spectrum

Having determined the conditions under which recoilless resonant absorption can occur, it is possible to devise a simple experiment to demonstrate the effect. A solid matrix containing the excited nuclei of a suitable isotope is used as the source of $\gamma$-rays. It is placed alongside a second matrix of identical material containing the same isotope in its ground state, which becomes the absorber. The intensity of the $\gamma$-radiation transmitted through the absorber is measured as a function of temperature. If there is no resonant absorption, the counting rate is independent of temperature. If resonant absorption occurs, then there should be a decrease in the transmission as the temperature is lowered, and the recoilless fraction increases. The resonantly absorbed $\gamma$-rays are effectively lost from the transmitted beam of radiation. Some are re-emitted in a random direction (i.e. resonant fluorescence), while others are lost by an internal conversion process during decay of the excited level (see p.11).

Although this was the experiment first used by R. L. Mössbauer, it is not particularly useful. The method which he also pioneered and has now been adopted exclusively is far more subtle. The definition of the energy of the $\gamma$-ray emitted by the source in a recoilless Mössbauer event is about 1 in $10^{12}$–$10^{13}$, and corresponds conveniently to the Doppler energy shifts produced by small movements. If the source and absorber are in relative motion with a velocity $v$, then the effective value of $E_\gamma$ 'seen' by the absorber differs from the true energy by a small Doppler shift energy of $\epsilon = (v/c)E_\gamma$. In the event that $v$ is zero, the emission and absorption profiles completely overlap and the absorption is at a maximum. Any increase or decrease in $v$ can only decrease the overlap. If $v$ is very large (of either sign) there will be no overlap and no absorption. It therefore follows that a record of transmission as a function of the velocity $v$ will show an 'absorption spectrum'. This is illustrated schematically in Fig.1.2. By convention a positive velocity is taken to be a closing velocity as this represents an increase in the apparent energy of the photon arriving at the absorber.

The lineshape of the absorption is derived simply from the Heisenberg uncertainty in the energy. The distribution of the recoilless source radiation is given by

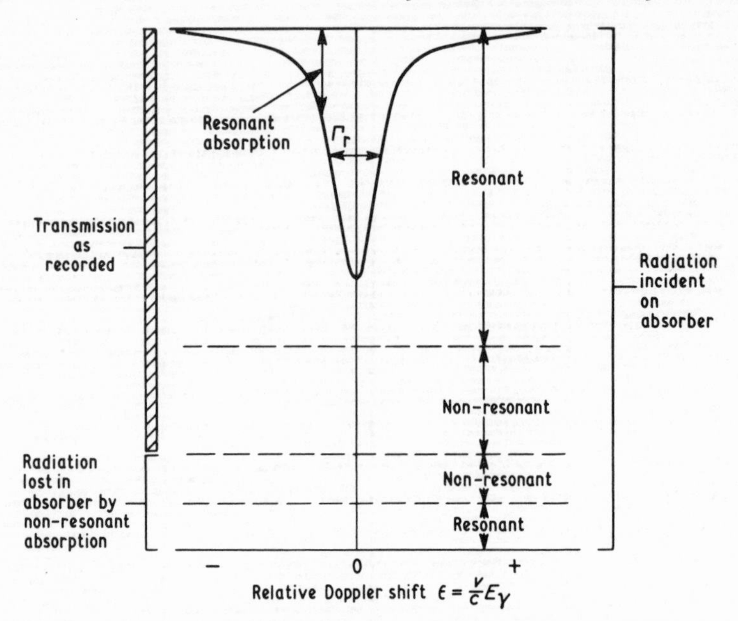


Fig. 1.2 A schematic representation of a Mössbauer transmission spectrum produced by Doppler scanning.

$$N(E)\mathrm{d}E = \frac{f\Gamma}{2\pi}\frac{\mathrm{d}E}{(E - E_\gamma)^2 + (\Gamma/2)^2} \tag{1.12}$$

where $N(E)$ is the probability of the energy being between $E$ and $E + \mathrm{d}E$, $f$ is the recoilless fraction of the source and $\Gamma$ is the Heisenberg width. This function which is usually referred to as the Lorentzian distribution has a maximum value at $E = E_\gamma$. The cross-section for resonant absorption, $\sigma(E)$, is similarly expressed [6, 7] as

$$\sigma(E) = \sigma_0 \frac{(\Gamma/2)^2}{(E - E_\gamma)^2 + (\Gamma/2)^2} \tag{1.13}$$

where $\sigma_0$ is a nuclear constant called the absorption cross-section given by

$$\sigma_0 = 2\pi(\hbar c/E_\gamma)^2 \frac{2I_\mathrm{e} + 1}{2I_\mathrm{g} + 1}\frac{1}{1 + \alpha} \tag{1.14}$$

where $I_e$ and $I_g$ are the nuclear spin quantum numbers of the excited and ground states and $\alpha$ is the internal conversion coefficient of the $\gamma$-ray [Note: $\gamma$-emission does not always lead to an observable $\gamma$-ray. In a proportion of events an atomic *s*-electron is ejected instead in a process known as internal conversion. $\alpha$ is defined as the ratio of the number of conversion electrons to the number of $\gamma$-ray photons emitted. Values for $\sigma_0$ are often quoted in units of barns (1 barn = $10^{-24}$ cm$^2$ = $10^{-28}$ m$^2$)]. For $\sigma_0$ to be large, both $E_\gamma$ and $\alpha$ should be small.

In any real absorber of finite thickness the resonant absorption is in direct competition with other non-resonant processes such as Compton scattering. A general evaluation of the integrated absorption intensity or 'transmission integral' is extremely difficult, but to a first approximation the basic behaviour is comparatively simple [8]. The recorded absorption spectrum for infinitely thin sources and absorbers has a Lorentzian distribution with a width at half-height of $\Gamma_r = 2\Gamma$. For an absorber of finite thickness containing $t_a$ resonant atoms per unit area of cross-section and with a recoilless fraction of $f_a$, an 'effective thickness' is usually defined by the dimensionless parameter $T_a = f_a\sigma_0 t_a$. The absorption shape approximates very closely to Lorentzian but with a broadened width $\Gamma_r$ given by

$$\Gamma_r/\Gamma \simeq 2 + 0.27T_a \tag{1.15}$$

provided that $T_a < 5$; i.e. the line broadens with an initially linear dependence on thickness. Self-resonance in the source matrix results in an additional line broadening which is approximately independent of absorber thickness.

The probability that any given photon will be resonantly absorbed can be enhanced by increasing the thickness $T_a$. However, the intensity of the incident $\gamma$-ray beam is also strongly attenuated by conventional non-resonant absorption processes, and this places an upper limit on the value of $T_a$ which can be used effectively. The absorption may be defined in terms of the transmitted intensity at the resonant maximum ($I_0$) and the transmitted intensity at a large Doppler velocity where the absorption is zero ($I_\infty$) by

$$A = (I_\infty - I_0)/I_\infty \tag{1.16}$$

It has been evaluated [8] to be

$$A = f[1 - e^{-T_a/2} J_0(iT_a/2)] \tag{1.17}$$

where $f$ is the recoilless fraction of the source and $J_0(x)$ is a zero-order Bessel function. The value of $A/f$ as a function of $T_a$ is plotted in Fig.1.3, and illustrates how the absorption shows a saturation behaviour with increasing thickness. The effects of non-resonant attenuation combined with more practical problems concerning background radiation levels make it advisable to use the smallest value of $T_a$ which gives an adequate absorption. In practice there is usually an optimum range of absorber thickness for each isotope which is quickly derived from experience.

One of the unfortunate problems which arises is that it is difficult to determine absolute values for the recoilless fractions $f$ and $f_a$ directly from the absorption intensity without tedious and difficult experimentation. However, in the majority of applications accurate values for these parameters are not required.

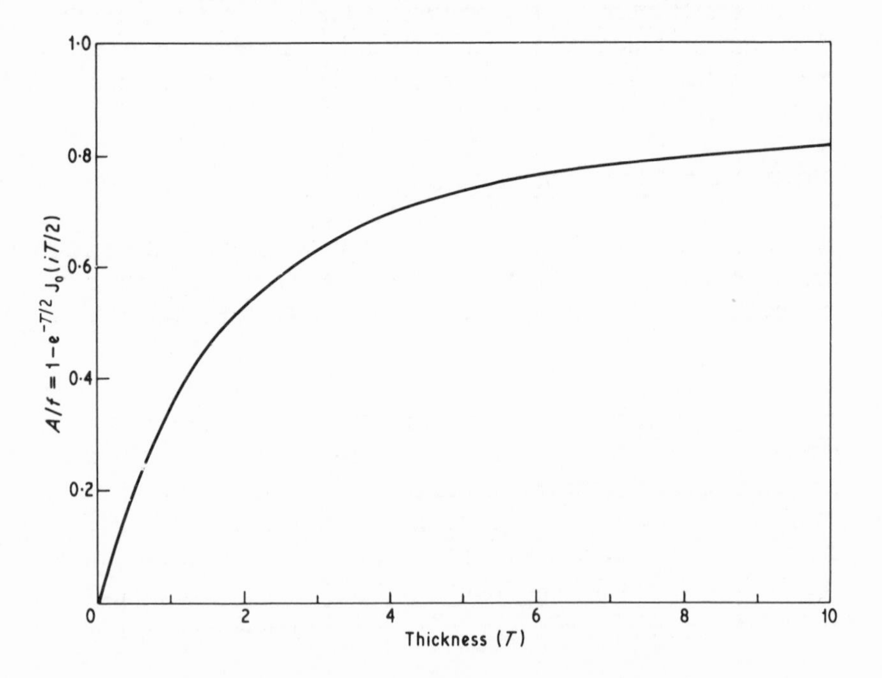


Fig. 1.3 The function $A/f = 1 - e^{-T/2} J_0(iT/2)$ showing how the absorption, $A$, shows a saturation behaviour with increasing thickness, $T$.

## 1.4 The Mössbauer spectrometer

The Mössbauer spectrum is a record of the transmission of resonant $\gamma$-rays through an absorber as a function of the Doppler velocity with respect to the source. It is therefore quite simply a record of transmission as a function of the energy of the incident radiation. The major difference from other forms of transmission spectroscopy is that the hazard to health in using high-energy radiation, combined with the high cost of producing the radioisotopes, necessitates the use of relatively low photon flux-densities. In consequence, much longer counting times are required to achieve a given resolution (hours rather than minutes). The statistics of the random counting predicts that the standard deviation in $N$ registered events is $\sqrt{N}$: thus the standard deviation in 10 000 counts is 1% and in 1 000 000 counts is 0.1% of the total count. A prospective gain in resolution must be balanced against the longer experimental time involved, which brings with it the problem of long-term reproducibility in the measuring equipment. To double the resolution requires four times the counting time, and makes it imperative to obtain as large a percentage absorption as possible.

The measurement of a Mössbauer spectrum is almost exclusively carried out by repetitively scanning the whole velocity range required, thereby accumulating the whole spectrum simultaneously, and allowing continuous monitoring of the resolution. This can be achieved electromechanically, and a typical modern Mössbauer spectrometer is illustrated schematically in Fig.1.4. The following description is a brief outline of the major principles involved. For a more detailed account of instrumentation and experimental methods the reader is referred to a recent review by Kalvius and Kankeleit [9].

The major component is a device known as a multichannel analyser which can store an accumulated total of $\gamma$-counts, using binary memory storage like a computer, in one of several hundred individual registers known as channels. Each channel is held open in turn for a short time interval of fixed length, which is derived from a very stable constant-frequency clock device. Any $\gamma$-counts registered by the detection system during that time interval are added to the accumulated total already stored in the channel. The sequential accessing of the channels is completed in about 1/20 of a second, and is repeated *ad infinitum.*

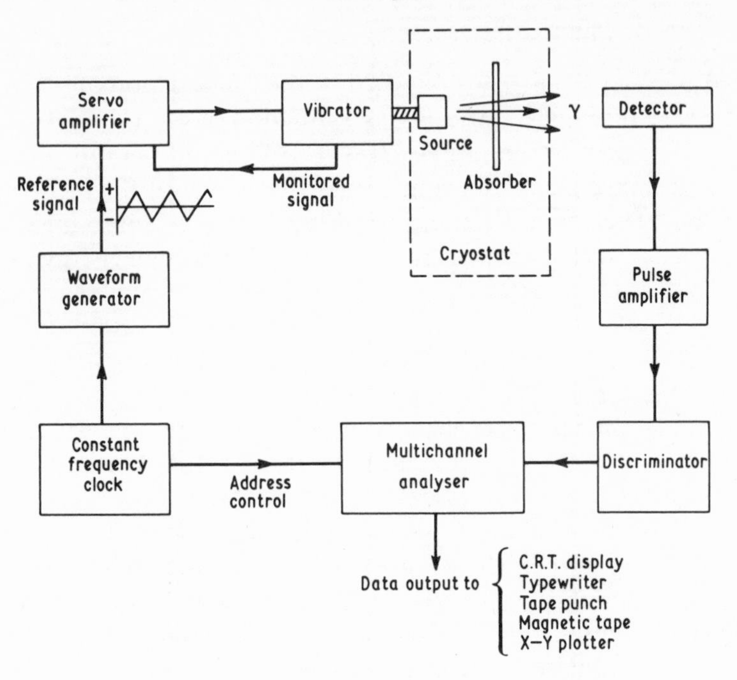


Fig. 1.4 A schematic arrangement for a Mössbauer spectrometer.

The timing pulses from the clock may also be used to synchronize a voltage waveform, which is used as a command signal to the servo-amplifier controlling an electromechanical vibrator. The latter moves the source relative to the absorber. A waveform, with a voltage increasing linearly with time, imparts a motion with a constant acceleration in which the drive shaft and the source spend equal time intervals at each velocity increment. The multichannel analyser and the drive are synchronized so that the velocity changes linearly from $-v$ to $+v$ with increasing channel number. In this way the source is always moving at the same velocity when a given channel is open.

The $\gamma$-ray detection system is conventional. A scintillation counter, gas proportional counter, or lithium-drifted germanium detector may be used according to circumstances [10]. The pulses from the detector are amplified, passed through a discriminator which rejects most of the non-resonant background radiation, and finally are fed to the open channel address.

The geometric arrangement of the source, absorber and detector is important. A $\gamma$-ray emitted along the axis of motion of the source has a Doppler shift relative to the absorber of $E_\gamma v/c$, but any $\gamma$-ray which travels to the detector along a path at an angle $\theta$ to this axis has an effective Doppler shift of only $E_\gamma v \cos\theta/c$. This 'cosine-effect' of solid-angle will therefore cause a spread in the apparent Doppler energy of the $\gamma$-rays and hence line broadening [11]. The effect can be reduced by maintaining an adequate separation between the source and detector, or by collimation.

The velocity range required to completely encompass the absorption is usually less than $\pm 10$ mm $s^{-1}$, so that the overall displacement of the source during a scan is barely discernible. Absolute calibration of the drive is not easy, but can be achieved by mounting a diffraction grating or mirror on the shaft as part of an interferometer. For most purposes it is more convenient to rely on an internal standard. As will be seen in the next chapter, it is possible for an absorber to show more than one resolved absorption line. If such a material is chemically reproducible and stable, and the lines are measured on an instrument with absolute calibration, it can then be used in other laboratories to establish the velocity amplitude and linearity of uncalibrated instruments. For example, it is common practice to use a magnetic $\alpha$-iron foil which has six lines as a calibrant for $^{57}Fe$ work, and to quote all measurements of velocity relative to the centroid of this reference spectrum.

Very few Mössbauer resonances are easy to record at room temperature, the main exceptions being $^{57}Fe$ and $^{119}Sn$, and in many instances it is necessary to cool at least the absorber, and sometimes the source as well, to increase the recoilless fractions. Because the source is moving, this is not without its difficulties. There is also the possibility that some of the hyperfine interactions (to be described in Chapter 2) are temperature dependent. For this reason alone, low-temperature measurements have become popular [9]. Numerous commercial cryostats are now available, the commonly used refrigerants being liquid nitrogen for temperatures down to 78 K and liquid helium down to 4.2 K. Cryostats with a fully-variable temperature control are also used, particularly in the study of phase transitions. In some applications it is desirable to apply a large external magnetic field to the absorber, and superconducting magnet installations are available which can produce magnetic flux densities of up to $10T$.

Although this discussion has been weighted in favour of using a transmission experiment to observe resonant absorption, it is also possible to detect resonant fluorescence by a scattering experiment. The resonant absorption event may be recorded by detecting the scattered $\gamma$-ray on de-excitation (alternatively, if the internal conversion coefficient $\alpha$ is large, the conversion-electron or X-ray produced may be registered). This has the advantage that non-resonant radiation from the source is not recorded, and the major remaining contributions to the radiation background are scattered higher-energy $\gamma$-rays and non-resonant Rayleigh scattering. The main disadvantages are that the intensity of the scattered radiation is very weak, and the large solid geometry requirement for a high count rate results in broadening of the Mössbauer line. A further problem, which may arise if scattered $\gamma$-rays are counted, is the possibility of anomalous effects dependent on the scattering angle. These are due to interference between resonant Mössbauer scattering and the non-resonant Rayleigh scattering [12] and to diffraction [13].

The principal application of scattering experiments is in the study of surface phenomena, and examples of this are given in Chapter 9.

## 1.5 Mössbauer isotopes

Several requirements can be formulated which must be fulfilled if there is to be an easily observable Mössbauer resonance:

(1) The energy of the $\gamma$-ray must be between 10 and 150 keV, preferably less than 50 keV, because the recoilless fraction $f$ and resonant cross-section $\sigma_0$ both decrease as $E_\gamma$ increases. This is the main reason why no resonances are known for isotopes lighter than $^{40}K$. The $\gamma$-transitions in light nuclei are usually very energetic.

(2) The half-life of the first excited state which determines $\Gamma$ should be between about 1 and 100 ns. If $t_{1/2}$ is very long, then $\Gamma$ is so narrow that mechanical vibrations destroy the resonance condition, and if $t_{1/2}$ is very short then $\Gamma$ will probably be so broad as to obscure any useful hyperfine effects.

(3) The internal conversion coefficient $\alpha$ should be small ($<10$) so that there is a good probability of detecting the $\gamma$-ray.

(4) A long-lived precursor should exist which can populate the required excited level. This usually means an isotope which decays by $\beta$-decay, electron capture, or isomeric transition. Although

some experiments have been done with energetic nuclear reactions *in situ* such as Coulombic excitation or (d, p) reactions, these methods are not applicable to routine measurements.

(5) The ground-state isotope should be stable and have a high natural abundance so that isotopic enrichment of absorbers is unnecessary.

Despite these restrictions, a Mössbauer resonance has been recorded in 100 transitions of 83 different isotopes in 44 elements, the majority of the heavy elements in fact. These are listed at the end of the book. However, many of these resonances can only be recorded with difficulty at the present time. Some of the more useful resonances, many of which are illustrated with examples in later chapters, are listed in Table 1.2. Some of their nuclear properties are also given, including the nuclear spin states $I_g$ and $I_e$ and their parity, and the internal conversion coefficient $\alpha$.

The most popular Mössbauer resonance is undoubtedly the 14.41-keV $\gamma$-transition in $^{57}$Fe, and the decay scheme for the $^{57}$Co parent is shown in Fig.1.5, together with the decay schemes for the $^{119}$Sn, $^{129}$I and $^{193}$Ir resonances. The conveniently long-lived $^{57}$Co nucleus decays by electron-capture with high efficiency to the 136.32-keV level of $^{57}$Fe, from whence 85% of the decays proceed to the 14.41-keV level. The lifetime of this state is 99.3 ns, giving a Heisenberg width of $\Gamma_r = 0.192$ mm s$^{-1}$ in the Mössbauer resonance. The large value of $\alpha$ for this level (8.17) unfortunately reduces the source efficiency to less than 10%, but the $\gamma$-ray is well resolved from the two higher energy $\gamma$-rays and the 7-keV X-rays produced by internal conversion. Although the natural abundance of $^{57}$Fe is only 2.19%, the absorption cross-section of $\sigma_0 = 2.57 \times 10^{-18}$ cm$^2$ is unusually large, and results in a satisfactory resonance at room temperature. Metals and refractory materials can give an acceptable absorption even above 1000 K, although organometallic compounds with a low effective Debye temperature are more effective when cooled.

One of the most important experimental aspects is in the choice of a source matrix. It is very desirable to have a high recoilless fraction and a single emission line unbroadened by hyperfine interactions. The best choice of host matrix is usually a high-melting metal or a refractory oxide of cubic structure; for example the most popular host for $^{119m}$Sn is the oxide $BaSnO_3$, and $^{57}$Co is usually diffused into a metal, such as palladium, platinum or rho-

Table 1.2 Nuclear parameters for selected Mössbauer transitions

| Isotope | $E_\gamma$/keV | $\Gamma_r$/ (mm s$^{-1}$) | $I_g$ | $I_e$ | $\alpha$ | Natural abundance (%) | Nuclear decay* |
|---|---|---|---|---|---|---|---|
| $^{57}$Fe | 14.41 | 0.192 | 1/2− | 3/2− | 8.17 | 2.17 | $^{57}$Co(EC 270 d) |
| $^{61}$Ni | 67.40 | 0.78 | 3/2− | 5/2− | 0.12 | 1.25 | $^{61}$Co($\beta^-$99 m) |
| $^{99}$Ru | 90 | 0.147 | 5/2+ | 3/2+ | – | 12.63 | $^{99}$Rh(EC 16 d) |
| $^{119}$Sn | 23.87 | 0.626 | 1/2+ | 3/2+ | 5.12 | 8.58 | $^{119m}$Sn(IT 250 d) |
| $^{121}$Sb | 37.15 | 2.1 | 5/2+ | 7/2+ | ~10 | 57.25 | $^{121m}$Sn($\beta^-$ 76 y) |
| $^{125}$Te | 35.48 | 5.02 | 1/2+ | 3/2+ | 12.7 | 6.99 | $^{125}$I(EC 60 d) |
| $^{127}$I | 57.60 | 2.54 | 5/2+ | 7/2+ | 3.70 | 100 | $^{127m}$Te ($\beta^-$109 d) |
| $^{129}$I | 27.72 | 0.59 | 7/2+ | 5/2+ | 5.3 | nil | $^{129m}$Te ($\beta^-$33 d) |
| $^{129}$Xe | 39.58 | 6.85 | 1/2+ | 3/2+ | 11.8 | 26.44 | $^{129}$I($\beta^-$ 1.7 x 10$^7$ y) |
| $^{149}$Sm | 22.5 | 1.60 | 7/2− | 5/2− | ~12 | 13.9 | $^{149}$Eu(EC 106 d) |
| $^{151}$Eu | 21.6 | 1.44 | 5/2+ | 7/2+ | 29 | 47.8 | $^{151}$Gd(EC 120 d) |
| $^{161}$Dy | 25.65 | 0.37 | 5/2+ | 5/2− | ~2.5 | 18.88 | $^{161}$Tb($\beta^-$ 6.9 d) |
| $^{169}$Tm | 8.40 | 9.3 | 1/2+ | 3/2+ | 220 | 100 | $^{169}$Er($\beta^-$ 9.4 d) |
| $^{182}$W | 100.10 | 2.00 | 0+ | 2+ | 3.2 | 26.4 | $^{182}$Ta($\beta^-$ 115 d) |
| $^{189}$Os | 69.59 | 2.41 | 3/2− | 5/2− | 8.2 | 16.1 | $^{189}$Ir(EC 13.3 d) |
| $^{193}$Ir | 73.0 | 0.60 | 3/2+ | 1/2+ | ~6 | 61.5 | $^{193}$Os($\beta^-$ 31 h) |
| $^{197}$Au | 77.34 | 1.87 | 3/2+ | 1/2+ | 4.0 | 100 | $^{197}$Pt($\beta^-$18 h) |
| $^{237}$Np | 59.54 | 0.067 | 5/2+ | 5/2− | 1.06 | nil | $^{241}$Am ($\alpha$458 y) |

*EC = electron capture, $\beta^-$ = beta-decay, IT = isomeric transition, $\alpha$ = alpha-decay

dium. However, these were chosen as the result of experience, and the development of a good source is of prime importance in studying a new resonance.

In some instances where the absorption is too weak, a substantial improvement can be obtained by using isotopic enrichment.

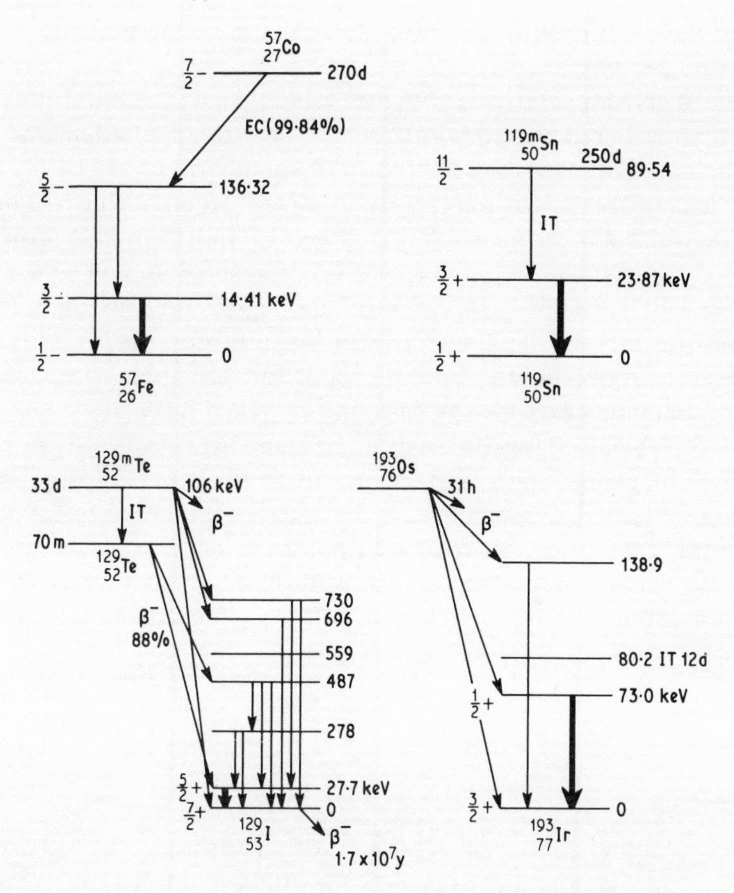


Fig. 1.5 The nuclear decay schemes for the Mössbauer resonances in $^{57}$Fe, $^{119}$Sn, $^{129}$I and $^{193}$Ir.

Thus $^{57}$Fe has an abundance of only 2.17% in natural iron, but material enriched to 90% can be used in compounds with a very low total iron content (eg. dilute iron alloys or haeme-proteins) to obtain an adequate cross-section for absorption.

## 1.6 Computation of data

The characteristic digital data comprising a Mössbauer spectrum are in an ideal form for statistical calculations by digital computer. For this reason the multichannel analyser is usually equipped to output data in a form suitable for input to one of the standard computer peripherals.

Each spectrum comprises $N$ digital values, $Y_i$, of the number of $\gamma$-counts registered at velocities $x_i$. The statistics of spontaneous radioactive decay processes follow a Poisson distribution [14]. If the mean value of the number of counts recorded in a given time is $Y$, then the standard deviation in $Y$ is $\sqrt{Y}$. In its simplest form the Mössbauer spectrum comprises a single absorption line which has a Lorentzian shape, and can therefore be specified completely by four parameters; the linewidth, the line position, the intensity of the absorption, and the baseline count for zero absorption. These parameters may always be found by visual inspection, but more precise values, together with their standard deviations, are obtained by computing a least-squares fit to the data. Appropriate starting values of the $n$ parameters are 'guessed' and used to calculate the 'theoretical' value of data point i to be $A_i(n)$. The discrepancy between the observed data and the assumed model can be represented by the weighted sum of the squares of the differences

$$F = \sum_{i=1}^{N} \left\{ \frac{[Y_i - A_i(n)]^2}{Y_i} \right\} \tag{1.18}$$

This expression can be used as a 'goodness-of-fit' function and minimized by standard mathematical routines to give the best values of the $n$ variables. The function $F$ has $N$-$n$ degrees of freedom and the properties of a chi-squared distribution at the minimum [14]; it may be related to the probability that the theoretical model is a valid description of the data, and in conjunction with statistical tables [15] its value indicates the degree of confidence to be placed on the final answers. For example, a computed fit to data with 400 degrees of freedom is within the 25–75% confidence limits of the $\chi^2$ distribution if $380 < F < 419$, and within the 5–95% limits if $355 < F < 448$. A value outside these limits is generally indicative of either faulty data or an inappropriate theoretical model.

The most commonly used theoretical function is a summation of Lorentzian lines, but the same basic principles are valid for any function which correctly describes the data. However, one word of caution must be given. A good fit to data is *not* unambiguous proof that the theoretical function is the correct one. In a complicated spectrum it is quite feasible to fit a function which has no physical significance. The final data analysis *must* be compatible with other scientific evidence, and inevitably there will be instances where it is not possible to distinguish between alternative hypotheses.

## References

[1] Mössbauer, R. L. (1958) *Z. Physik,* **151**, 124.
[2] Kuhn, W. (1929) *Phil. Mag.,* **8**, 625.
[3] Metzger, F. R. (1959) *Progr. Nuclear Phys.,* **7**, 54.
[4] Lipkin, H. J. (1960) *Ann. Phys.,* **9**, 332.
[5] Petzold, J. (1961) *Z. Physik,* **163**, 71; Wisscher, W. M. (1960) *Ann. Phys.,* **9**, 194.
[6] Frauenfelder, H. (1962) *The Mössbauer Effect,* W. A. Benjamin Inc., N.Y.
[7] Jackson, J. D. (1955) *Can. J. Phys.,* **33**, 575.
[8] Margulies, S. and Ehrman, J. R. (1961) *Nuclear Instr. Methods,* **12**, 131.
[9] Kalvius, G. M. and Kankeleit, E. (1972) Recent improvements in instrumentation and methods of Mössbauer spectroscopy. In *Mössbauer Spectroscopy and its Applications,* International Atomic Energy Agency, Vienna, p.9.
[10] (1965) *Alpha-, beta-, and gamma-ray spectroscopy* Vol.1, ed. K. Siegbahn, North-Holland, Amsterdam.
[11] Riesenman, R., Steger, J. and Kostiner, E. (1969) *Nuclear Instr. Methods,* **72**, 109.
[12] Black, P. J., Longworth, G. and O'Connor, D. A. (1964) *Proc. Phys. Soc.,* **83**, 925.
[13] Black, P. J. and Duerdoth, I. P. (1965) *Proc. Phys. Soc.,* **84**, 169.
[14] Mulholland, H. and Jones, C. R. (1968) *Fundamentals of Statistics,* Butterworths, London.
[15] Beyer, W. H. (editor), (1968) *Handbook of Tables for Probability and Statistics,* 2nd edition. The Chemical Rubber Co., Cleveland.

CHAPTER TWO

# *Hyperfine Interactions*

In Chapter 1 it was shown that the Mössbauer spectrum is a record of the intensity of a transmitted $\gamma$-ray beam as a function of the Doppler velocity between the source and absorber. The resonant absorption line has a Lorentzian shape of width $\Gamma_r$ and is centred at zero relative velocity between source and absorber. The practical application of this measurement would be limited were it not for the fact that the energies of the nuclear states are weakly influenced by the chemical environment. It is possible to detect these extremely small effects because of the high definition (better than 1 in $10^{12}$) of the $\gamma$-ray energy.

There are three principal interactions to consider. These are:

(1) A change in the electric monopole or Coulombic interaction between the electronic and nuclear charges, which is caused by a difference in the size of the nucleus in its ground and excited states. This is seen as a shift of the absorption line away from zero velocity and is variously known as the chemical isomer shift, isomer shift, or centre shift, and designated by the symbol $\delta$.

(2) A magnetic dipole interaction between the magnetic moment of the nucleus and a magnetic field. The origin of the latter may be intrinsic or extrinsic to the atom. The result is a multiplet line structure in the spectrum known as magnetic hyperfine splitting.

(3) An electric quadrupole interaction between the nuclear

quadrupole moment and the local electric field gradient tensor at the nucleus. This also results in a multiple-line spectrum.

Magnetic hyperfine interactions were first observed by Pound and Rebka in 1959 [1], and the chemical isomer shift and quadrupole interactions by Kistner and Sunyar in 1960 [2]. At first sight the rapid development of Mössbauer spectroscopy from then onwards seems extraordinary, but it should be realized that most of the underlying theory was already available. The chemical isomer shift is similar in origin to the isotope shift in atomic spectra. Electric quadrupole interactions are the basis of nuclear quadrupole resonance, while the magnetic dipole interactions together with the theory of time-dependent processes were familiar in nuclear magnetic resonance and electron spin resonance spectroscopy.

All three effects may occur together, but only the magnetic and quadrupole interactions are directional and thus have a complicated interrelationship. The chemical isomer shift behaves independently of the other two interactions, and is conveniently considered first.

Although this chapter emphasizes the ways in which the hyperfine interactions of the nucleus are used to investigate the chemical environment, it should be noted that many of the early measurements on each Mössbauer resonance were made to obtain values for the nuclear parameters. For example, although the spin, magnetic moment, and quadrupole moment are usually known for the ground state of the nucleus, this is not necessarily the case for an excited state. It has often been necessary to determine one or more of these parameters from the hyperfine interactions in appropriately selected compounds.

## 2.1 The chemical isomer shift

In most instances it is adequate to consider the Coulombic interaction between the electrons and the nucleus as if the latter were a point charge. This assumption is made for example in solving the Schrödinger wave equation for the electronic structure of the atom. Such a simplification predicts that there will be no change in the Coulombic interaction energy when the nuclear excited state decays to the ground state. However, the nucleus does have a finite size which may change fractionally during the transition. It must also be realized that an *s*-electron wavefunction has a finite value *inside* the nuclear radius, and is directly responsible for the change in electrostatic energy observed.

The integrated Coulombic energy for an electron of charge $-e$ moving in the field of a point nucleus of charge $+Ze$ is given by

$$E_0 = -\frac{Ze^2}{4\pi\epsilon_0}\int_0^\infty \Psi^2 \frac{d\tau}{r} \tag{2.1}$$

where $\epsilon_0$ is the permittivity of a vacuum, $r$ is the radial distance, and $-e\Psi^2$ is the charge density of the electron in volume element $d\tau$. If the nucleus is spherical with a radius $R$, then equation (2.1) is valid for $r > R$, but overestimates $E_0$ for $r < R$. It is possible to calculate an approximate correction, $W$, by assuming a model for the proton charge density within the nucleus [3]. If this is taken to be uniform, and certain simplifying approximations are made in the mathematics, the comparatively simple expression obtained is

$$W = \frac{1}{10\epsilon_0} Ze^2R^2|\Psi_s(0)|^2 \tag{2.2}$$

where $|\Psi_s(0)|^2$ is the non-relativistic Schrödinger wavefunction at $r = 0$.

If the nucleus changes its radius by a small increment $\delta R$ during its transition from the excited state to the ground state, there will be a simultaneous change in electrostatic energy given by

$$\Delta W = \frac{1}{5\epsilon_0} Ze^2R^2 \frac{\delta R}{R} |\Psi_s(0)|^2 \tag{2.3}$$

The value of $\delta R/R$ is characteristic of each transition, but is typically of the order of $10^{-4}$ and may be of either sign. A positive sign means that the nucleus shrinks on de-excitation. The Mössbauer experiment compares the difference in energy between the nuclear transitions in the source and absorber, so that the chemical isomer shift as observed is given by

$$\delta = \frac{1}{5\epsilon_0} Ze^2R^2 \frac{\delta R}{R} (|\Psi_s(0)_{\mathrm{absorber}}|^2 - |\Psi_s(0)_{\mathrm{source}}|^2) \tag{2.4}$$

This can then be related quite simply to the measured shift in Doppler velocity units, $v$, by $v = (c/E_\gamma)\delta$. (The customary symbol for the chemical isomer shift, $\delta$, is not to be confused with the change in nuclear radius, $\delta R$).

Equation (2.4) can be seen to be the product of a chemical term (the electron density at $r = 0$) and a nuclear term (the change in nuclear radius). The latter is a constant for a particular transition, and thus it is possible to study relative changes in electron density directly. Equations similar to (2.4) are sometimes used which express the nuclear radius as a mean-square value to indicate that it is not necessarily spherical, but these may be converted to the expression above by using the relationships $\delta\langle R^2\rangle/\langle R^2\rangle = 2\delta R/R$ and $\langle R_e^2\rangle - \langle R_g^2\rangle = \frac{6}{5}R^2(\delta R/R)$. For chemical applications it is often sufficient to compare chemical isomer shift values using the simplified expression

$$\delta = \text{constant} \times (|\Psi_s(0)_A|^2 - |\Psi_s(0)_B|^2) \tag{2.5}$$

where A and B are two different chemical environments of which B is either the source matrix or a reference absorber.

It is important to realize that $|\Psi_s(0)|^2$ is the $s$-electron density *at the nucleus,* and not the $s$-electron occupation in the formal chemical sense. If $\delta R/R$ is positive, a positive value of the chemical isomer shift, $\delta$, implies that the $s$-electron density at the nucleus in A is greater than in B. $|\Psi_s(0)|^2$ includes contributions from all the occupied $s$-electron orbitals in the atom, but is naturally more sensitive to changes which take place in the outer valence shells. Although the values of $|\Psi(0)|^2$ for $p$-, $d$- and $f$-electrons are zero, these orbitals nevertheless do have a significant *indirect* interaction with the nucleus via interpenetration shielding of the $s$-electrons. For example a $3d^54s^1$ configuration will have a larger value of $|\Psi_s(0)|^2$ than $3d^64s^1$ because in the latter case the extra $d$-electron shields the $4s$-electron from the nucleus.

In a number of instances a Mössbauer resonance can be observed for two or more transitions to the same ground state, or for several isotopes of the same element. The chemical isomer shifts of two different resonances in the same series of compounds should be interrelated purely by the difference in $\delta R/R$, and this appears to be the case. An interesting illustration is provided by the $^{127}I$ and $^{129}I$ spectra of $Na_3H_2IO_6$ shown in Fig.2.1. The source matrix was zinc telluride in each case but containing $^{127m}Te$ and $^{129m}Te$ respectively [4]. These isotopes $\beta$-decay to populate the 57.6-keV level of $^{127}I$ and the 27.7-keV level of $^{129}I$. The absorptions occur at positive and negative velocities because $\delta R/R$ is negative in $^{127}I$ and positive in $^{129}I$, while the *s*-electron density at the nucleus is greater in the source than the absorber. The relative magnitudes of the shift also differ because $[\delta R/R(^{127}I)]/[\delta R/R(^{129}I)] = -0.65$. The resonance lines have different widths because $\Gamma_r = 2.54$ mm s$^{-1}$ for $^{127}I$ and 0.59 mm s$^{-1}$ for $^{129}I$.

An accurate calibration of the chemical isomer shift scale for a particular isotope is not without difficulties. The change in nuclear radius, $\delta R/R$, cannot be determined independently of the chemical environment, which means that the electron density $|\Psi_s(0)|^2$, has to be estimated by molecular orbital methods in at least two compounds before a value for $\delta R/R$ can be obtained. Some of the problems which this creates are discussed in Chapter 4.

The observed line-shift is not entirely caused by the chemical isomer shift as described above. There is another generally smaller contribution termed the second-order Doppler shift, which was first observed by Pound and Rebka in 1960 [5]. The emitting or absorbing nucleus is not stationary but is vibrating on its lattice site. The period of vibration is much shorter than the Mössbauer lifetime so that the average displacement and velocity are effectively zero, but the mean-squared values of the velocity, $\langle v^2 \rangle$, are finite.

The relativistic equation for the Doppler effect on the apparent frequency $\nu$ of the emitted photon (as recorded at the absorbing nucleus) is

$$\nu = \nu_0 \left(1 - \frac{v}{c} \cos \alpha\right) \left(1 - \frac{v^2}{c^2}\right)^{-1/2} \tag{2.6}$$

where $\nu_0$ is the frequency for a stationary nucleus, and $v$ is the apparent relative velocity of the emitting nucleus along a direction making an angle $\alpha$ with the $\gamma$-ray direction (Note: $v$ is not the velocity along the direction of travel of the $\gamma$-ray). For an oscillatory

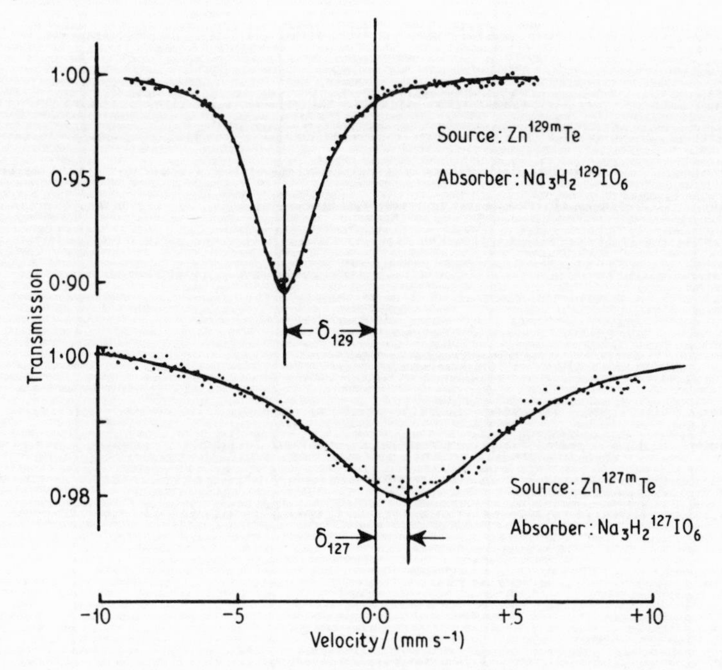


Fig. 2.1 The $^{127}$I and $^{129}$I spectra of $Na_3H_2IO_6$ showing the smaller line-width and larger chemical isomer shift of opposite sign for the latter ([4], Fig. 1)

motion the mean value $\langle v \rangle$ is zero so that only the second-order term in $\langle v^2 \rangle$ can influence the Mössbauer resonance. Hence for $v \ll c$

$$\nu \simeq \nu_0 \left(1 + \frac{\langle v^2 \rangle}{2c^2}\right) \tag{2.7}$$

This gives rise to a shift in the Mössbauer line of

$$\frac{\delta E}{E_\gamma} = \frac{\nu_0 - \nu}{\nu_0} = -\frac{\langle v^2 \rangle}{2c^2} \tag{2.8}$$

It is obvious that $\langle v^2 \rangle$ increases with rise in temperature. Accordingly, the Mössbauer resonance in an absorber moves to a more negative velocity as the temperature is raised. The second-order Doppler shift contribution to the observed chemical shift is smallest at absolute zero, but is still finite because of a zero-point motion of the nucleus.

The second-order Doppler shift is often ignored when comparing values of the total observed shift because it always acts in the same sense, but such a comparison should only be made for similar compounds at the same temperature, and small differences regarded as not significant. A more detailed discussion of the second-order Doppler shift will be found in Chapter 6.

## 2.2 Magnetic hyperfine interactions

The nucleus has a magnetic moment, $\mu$, when the spin quantum number, $I$, is greater than zero. Its energy is then affected by the presence of a magnetic field, and the interaction of $\mu$ with a magnetic flux density of $B$ is formally expressed by the Hamiltonian

$$\mathscr{H} = -\boldsymbol{\mu} \cdot \mathbf{B} = -g\mu_{\mathrm{N}}\mathbf{I} \cdot \mathbf{B} \tag{2.9}$$

where $\mu_{\mathrm{N}}$ is the nuclear magneton ($eh/4\pi m_{\mathrm{p}} = 5.04929 \times 10^{-27}$ A m$^2$ or J T$^{-1}$) and $g$ is the nuclear $g$-factor [$g = \mu/(I\mu_{\mathrm{N}})$]. Solving this Hamiltonian [6] gives the energy levels of the nucleus in the field to be

$$E_{\mathrm{m}} = -\frac{\mu B}{I} m_z = -g\mu_{\mathrm{N}} B m_z \tag{2.10}$$

where $m_z$ is the magnetic quantum number and can take the values $I, I-1, \ldots, -I$. In effect, the magnetic field splits the energy level into $2I + 1$ non-degenerate equi-spaced sublevels with a separation of $\mu B/I$.

In a Mössbauer experiment there may be a transition from a ground state with a spin quantum number $I_{\mathrm{g}}$ and magnetic moment $\mu_{\mathrm{g}}$ to an excited state with spin $I_{\mathrm{e}}$ and magnetic moment $\mu_{\mathrm{e}}$. In a magnetic field, both states will be split according to equations (2.9) and (2.10). Transitions can take place between sub-levels provided that the selection rule $\Delta m_z = 0, \pm 1$ is obeyed [this is called a magnetic dipole (M1) transition; for other selection rules see Section 2.5 on line intensities]. The resultant Mössbauer spectrum contains a number of resonance lines, but is nevertheless symmetrical about the centroid.

A typical example of magnetic hyperfine splitting is illustrated schematically in Fig.2.2, which is drawn to a scale appropriate to $^{119}$Sn. For this isotope $I_g = \frac{1}{2}$, $I_e = \frac{3}{2}$, $\mu_g = -1.041\mu_N$ and $\mu_e = +0.67\mu_N$. The change in sign of the magnetic moment results in a relative inversion of the multiplets. The six lines are the allowed $\Delta m_z = 0, \pm 1$ transitions, and the resultant spectrum is indicated in the stick diagram. The lines are not of equal intensity, but the 3:2:1:1:2:3 ratio shown here is often found for example in the $^{57}$Fe and $^{119}$Sn resonances in randomly oriented polycrystalline samples. A more detailed account of relative line intensities is given later in the chapter.

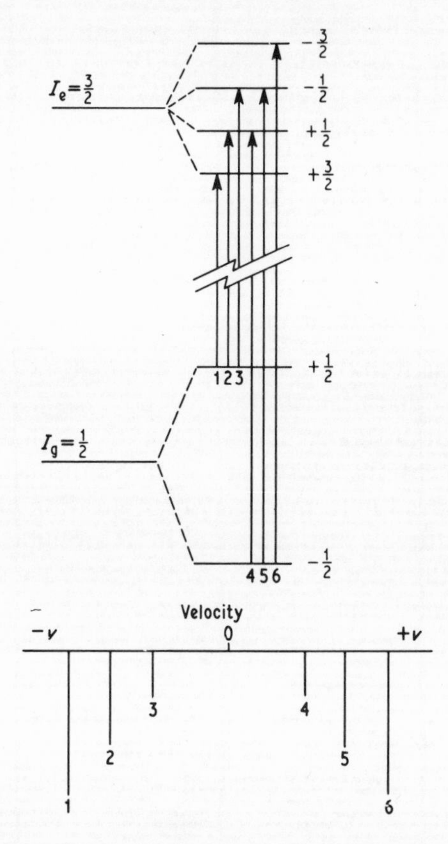


Fig. 2.2 The energy-level scheme and resultant spectrum for magnetic hyperfine splitting of an $I_g = \frac{1}{2} \rightarrow I_e = \frac{3}{2}$ transition. The relative splittings are scaled in accord with the magnetic moments of $^{119}$Sn; $\mu_g = -1.041\mu_N$ and $\mu_e = +0.67\mu_N$. The line intensity ratios of 3:2:1:1:2:3 are appropriate to a polycrystalline absorber.

The magnetic field may be applied by an external magnet, or it may be intrinsic to the compound in which case it is usually referred to as an 'internal' field. An unpaired electron in the atomic environment can induce an imbalance in electron spin-density at the nucleus, and thereby generate a local magnetic flux density which can be as large as 100 T. Examples of this may be found in Chapter 5.

The sign of an internal magnetic flux density, $B$, can sometimes be obtained by applying an additional externally generated magnetic flux density, $B_0$, with a large magnet. If the resultant flux density in the Mössbauer spectrum has increased to $B + B_0$ by parallel alignment of $B$ with $B_0$ (i.e. the internal field is parallel to the magnetization) then the sign is positive, whereas a field of $B - B_0$ signifies antiparallel alignment and a negative sign. This method fails if the magnetic anisotropy of the matrix is large enough to prevent rotation of the internal field into the direction of the applied field.

Time-dependent effects such as relaxation are discussed separately in Chapter 6.

## 2.3 Electric quadrupole interactions

The electric quadrupole interaction in Mössbauer spectroscopy is very similar to that in nuclear quadrupole resonance spectroscopy [7]. The main difference is that the latter is concerned with radiofrequency transitions *within* a hyperfine multiplet of a ground state nucleus, whereas the former is a $\gamma$-ray transition *between* the hyperfine multiplets of the nucleus in its ground and excited states. Only those nuclear states with $I > \frac{1}{2}$ have a nuclear quadrupole moment and hence show a quadrupole hyperfine splitting. In consequence it is often possible to observe a quadrupole interaction in the Mössbauer spectrum which derives from the excited state of the nucleus, even though the ground state has $I_g = 0$ or $I_g = \frac{1}{2}$ and therefore does not give an N.Q.R. signal.

The nuclear quadrupole moment, $Q$, is a measure of the deviation from spherical symmetry of the nuclear charge. It is expressed by

$$eQ = \int \rho r^2 (3\cos^2\theta - 1)\,d\tau \qquad (2.11)$$

where $+e$ is the charge on the proton, and $\rho$ is the charge density in the volume element $d\tau$ at a distance $r$ from the centre of the nucleus and at an angle $\theta$ to the axis of the nuclear spin. The magnitude of $Q$ is often referred to in units of barns (1 barn = $10^{-28}$ $m^2$). The sign of $Q$ can be positive or negative according to whether the nucleus is respectively elongated (prolate) or flattened (oblate) along the spin-axis.

The electrostatic potential at the nucleus due to a point charge $q$ at a distance $r$ is given by $V = q/(4\pi\epsilon_0 r)$ where $\epsilon_0$ is the permittivity of a vacuum. The interaction of the nuclear quadrupole moment with the electronic environment is expressed by the Hamiltonian

$$\mathscr{H} = -\frac{1}{6}e\mathbf{Q}\cdot\nabla\mathbf{E} \quad (2.12)$$

where $\nabla\mathbf{E}$ represents the electric field gradient at the nucleus (which is the derivative of the electric field, $E$, and hence the negative of the second derivative of the potential, $V$). $\nabla\mathbf{E}$ is a tensor quantity which can be written as

$$\nabla_i E_j = -\frac{\partial^2 V}{\partial x_i \partial x_j} = -V_{ij} \quad (2.13)$$

$$(x_i, x_j = x, y, z)$$

There are thus nine values of $V_{ij}$ to express $\nabla\mathbf{E}$ in a Cartesian axis system, $x, y, z$. A 'principal' axis system may always be defined such that all the $V_{ij}$ terms with $i \neq j$ are zero (the matrix of the tensor is diagonal), leaving the three finite 'principal' values $V_{xx}$, $V_{yy}$, and $V_{zz}$. Furthermore, $\Delta\mathbf{E}$ is a traceless tensor, so that

$$V_{xx} + V_{yy} + V_{zz} = 0 \quad (2.14)$$

As a result it is only necessary to specify two parameters to completely describe an electric field gradient tensor in its principal axis sytem. $V_{zz} = eq$ is taken to be the largest value of $|V_{ii}|$, and an asymmetry parameter, $\eta$, is defined by

$$\eta = (V_{xx} - V_{yy})/V_{zz} \quad (2.15)$$

such that $|V_{zz}| > |V_{yy}| \geqslant |V_{xx}|$ and $0 \leqslant \eta \leqslant 1$. The $z$ axis is then referred to as the 'major' axis, and $x$ is the 'minor' axis. The correct assignment of these axes is often obvious from the molecular geometry if the overall symmetry is high; e.g. in a molecule with axial symmetry, the $z$ axis is the symmetry axis. (Note: in an arbitrary axis sytem it is necessary to specify five parameters which may be $V_{xx}$, $V_{xy}$, $V_{zz}$, $V_{yy}$, and $V_{yz}$, or alternatively the principal values $V_{zz}$ and $\eta$ with three angles to specify their orientation with respect to the arbitrary $x$, $y$, $z$ axes.)

The Hamiltonian in equation (2.12) is more often written as

$$\mathscr{H} = \frac{eQ}{2I(2I-1)}[V_{zz}\hat{I}_z^2 + V_{yy}\hat{I}_y^2 + V_{xx}\hat{I}_x^2]$$

$$= \frac{e^2qQ}{4I(2I-1)}[3\hat{I}_z^2 - I(I+1) + \eta(\hat{I}_x^2 - \hat{I}_y^2)] \qquad (2.16)$$

where $\hat{I}_x$, $\hat{I}_y$ and $\hat{I}_z$ are quantum-mechanical spin operators. A completely general solution of this Hamiltonian is not possible, but exact expressions may be given under certain conditions.

If the electric field gradient tensor has axial symmetry ($\eta = 0$) the energy levels are given by

$$E_Q = \frac{e^2qQ}{4I(2I-1)}[3I_z^2 - I(I+1)] \qquad (2.17)$$

where the quantum number $I_z$ can take the $2I + 1$ values of $I, I-1, \ldots, -I$. If $I = \frac{3}{2}, \frac{5}{2}, \frac{7}{2}$ etc. the energy level is split into a series of Kramers' doublets with $\pm I_z$ states degenerate. An important example is $I = \frac{3}{2}$, which has two energy levels at $+e^2qQ/4$ for $I_z = \pm\frac{3}{2}$ and $-e^2qQ/4$ for $I_z = \pm\frac{1}{2}$. If the symmetry is lower than axial (i.e. when $\eta > 0$), an exact expression can only be given for $I = \frac{3}{2}$, and is

$$E_Q = \frac{e^2qQ}{4I(2I-1)}[3I_z^2 - I(I+1)](1 + \eta^2/3)^{1/2} \qquad (2.18)$$

with energy levels at $\pm(e^2qQ/4)(1 + \eta^2/3)^{1/2}$. Values for higher spin states must be calculated numerically.

A Mössbauer transition between states can take place according to the selection rule $\Delta m_z = 0, \pm 1$ for M1 dipole radiation. The most common example is the $I_g = \frac{1}{2} \rightarrow I_e = \frac{3}{2}$ transition in which the ground state with $I_z = \pm\frac{1}{2}$ is unsplit, and the excited state has two levels with $I_z = \pm\frac{3}{2}$ and $\pm\frac{1}{2}$ separated by $(e^2qQ/2)(1 + \eta^2/3)^{1/2}$. The resultant Mössbauer spectrum is a doublet with a separation called the quadrupole splitting of $\Delta = (e^2qQ/2)(1 + \eta^2/3)^{1/2}$. When $\eta = 0$, $\Delta = e^2qQ/2$ or half the quadrupole coupling constant defined in nuclear quadrupole resonance spectroscopy. Typical energy level schemes for $\frac{1}{2} \rightarrow \frac{3}{2}$ and $\frac{7}{2} \rightarrow \frac{5}{2}$ (scaled to $^{129}I$) transitions and the resultant spectra with relative line intensities for randomly oriented polycrystalline samples are illustrated in Fig.2.3.

The quadrupole spectrum is symmetrical only in the case of a $\frac{1}{2} \rightarrow \frac{3}{2}$ transition. In this particular instance there is no direct means of establishing the magnitude of $\eta$ or the sign of $V_{zz}$. However, as detailed later in the chapter, it is possible to determine these parameters, either from the angular dependence of the line intensities in single crystals, or from the spectrum observed when a large external magnetic field is applied.

Once $e^2qQ$ and $\eta$ have been measured, they may be related to the chemical environment of the resonant nucleus. The coupling constant $e^2qQ$ is the product of a nuclear constant ($eQ$) and the maximum value of $V_{ij}$ ($eq$). Considerable confusion is often caused by referring to the sign of $V_{zz} = eq$ without specifying whether $e$ is taken to be the charge of the electron or the proton. For this reason it is preferable to refer to the sign of $e^2qQ$ or of $q$.

If $I_g = 0$ or $\frac{1}{2}$ so that $Q_g = 0$, then the magnitude of $Q_e$ has to be established empirically by comparing the observed quadrupole splitting values with estimated values of $V_{zz}$. Fortunately, use of the quadrupole splitting is not entirely dependent on these intrinsically difficult calculations. The relative order of magnitude of $e^2qQ$ can be related comparatively easily to the valence orbitals of the atom. The numerical value of $V_{zz}$ due to an occupied electron orbital expressed by the wavefunction $\Psi$ is given by the integral

$$\frac{V_{zz}}{e} = q = -\frac{(1-R)}{4\pi\epsilon_0} \int \Psi^* \left( \frac{3\cos^2\theta - 1}{r^3} \right) \Psi \, d\tau \qquad (2.19)$$

where $(1 - R)$ is an empirical quantity less than unity, called the

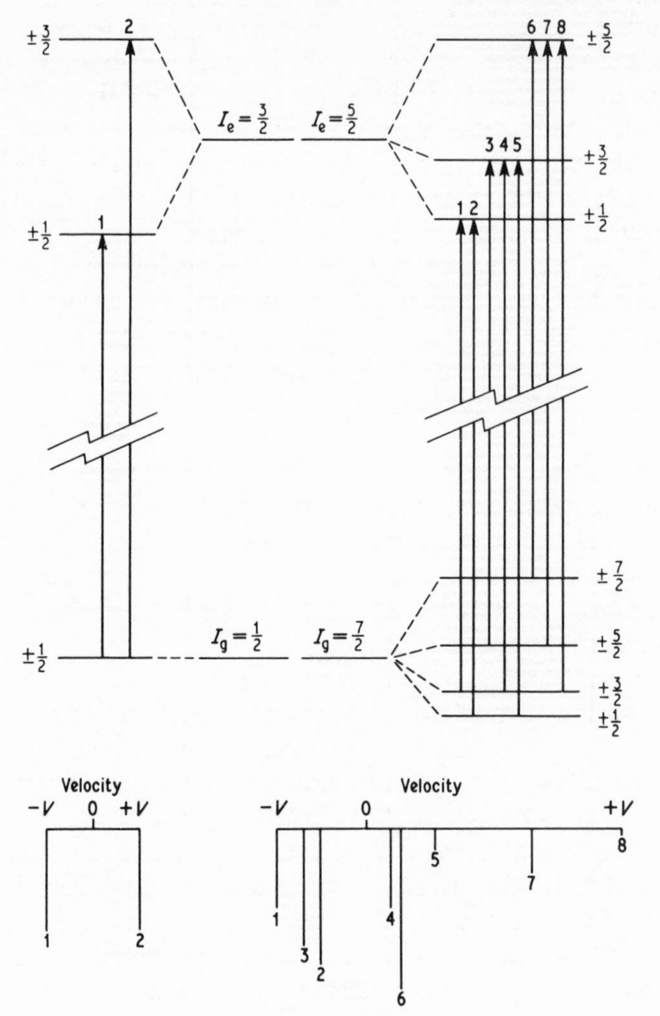


Fig. 2.3 The energy-level schemes and resultant spectra for quadrupole splitting of $I_g = \frac{1}{2} \rightarrow I_e = \frac{3}{2}$ and $I_g = \frac{7}{2} \rightarrow I_e = \frac{5}{2}$ (scaled to $^{129}$I) transitions.

Sternheimer shielding factor, which allows for an opposing polarization of the inner core electrons [8]. Similarly

$$\eta = \frac{1}{q} \int \Psi^* \left( \frac{3 \sin^2 \theta \cos 2\phi}{r^3} \right) \Psi \mathrm{d}\tau \qquad (2.20)$$

Table 2.1 The magnitude of $q$ and $\eta$ for the $p$- and $d$-orbitals

| Orbital | $4\pi\epsilon_0 q/(1-R)$ | $\eta$ |
|---|---|---|
| $p_z$ | $-\frac{4}{5}\langle r^{-3}\rangle$ | 0 |
| $p_x$ | $+\frac{2}{5}\langle r^{-3}\rangle$ | −3 |
| $p_y$ | $+\frac{2}{5}\langle r^{-3}\rangle$ | +3 |
| $d_{x^2-y^2}$ | $+\frac{4}{7}\langle r^{-3}\rangle$ | 0 |
| $d_{z^2}$ | $-\frac{4}{7}\langle r^{-3}\rangle$ | 0 |
| $d_{xy}$ | $+\frac{4}{7}\langle r^{-3}\rangle$ | 0 |
| $d_{xz}$ | $-\frac{2}{7}\langle r^{-3}\rangle$ | +3 |
| $d_{yz}$ | $-\frac{2}{7}\langle r^{-3}\rangle$ | −3 |

The magnitudes of $q$ and $\eta$, expressed in units of the expectation value of $1/r^3$, are given in Table 2.1 for the conventional $p$- and $d$-orbitals.

The total value of $q$ for a completely filled or half-filled shell of electrons is zero. An excess of electrons along the $z$ axis (occupying $p_z$, $d_z^2$, $d_{xz}$ or $d_{yz}$) results in a negative value of $q$, while an excess in the $xy$ plane ($p_x$, $p_y$, $d_{xy}$ or $d_{x^2-y^2}$) gives a positive value of $q$. The value of $\langle r^{-3}\rangle$ for a $4p$-electron will be significantly smaller than that of a $3p$-electron because of the greater radial extent of the former, and similarly a $3d$-electron will give a smaller value than a $3p$-electron, but probably greater than a $4p$-electron. The spherically symmetric $s$-electron gives no electric field gradient. The sign and magnitude of $e^2qQ$ can therefore be used empirically to measure the asymmetric occupation of the atomic valence orbitals, and many examples of this will be found in later chapters.

In some compounds the Mössbauer atom has an intrinsically high symmetry (e.g. the $Fe^{3+}$ $d^5$ ion has a half-filled shell and is a spherical $S$-state ion) but may still show a quadrupole splitting. The latter originates from charges external to the atom, such as other ions, which polarize the spherical inner core and can induce a very large electric field gradient at the nucleus. This 'lattice' contribution summed over individual charges $Z_i e$ may be written as

$$\frac{V_{zz}}{e} = q_{\text{latt}} = \frac{(1-\gamma_\infty)}{4\pi\epsilon_0} \sum_i Z_i \frac{3\cos^2\theta - 1}{r_i^3} \tag{2.21}$$

where $\gamma_\infty$ is the Sternheimer antishielding factor [8]. This parameter expresses the core polarization and is usually large and negative. The lattice contribution also exists when the valence orbitals are asymmetrically occupied, but is usually acknowledged to be a minority contribution unless the co-ordinate geometry is very distorted. Formal attempts to subdivide $q$ into valence and lattice terms are generally arbitrary and not altogether convincing.

## 2.4 Combined magnetic and quadrupole interactions

Both the magnetic and quadrupole hyperfine interactions express a directional interaction of the nucleus with its environment. However, when the two are present together, their respective principal axes are not necessarily co-linear, and it is not surprising therefore to find that the resultant behaviour can be much more complex. The formal Hamiltonian which is the sum of equations (2.9) and (2.12) has no general solution, but the predicted spectrum may always be obtained by numerical computation.

One of the few useful restricted solutions for a $\frac{1}{2} \rightarrow \frac{3}{2}$ transition is the case where the quadrupole interaction is very much weaker than the magnetic term, and can be treated as a small perturbation upon the latter. The resultant energy levels are given by

$$E_{QM} = -g\mu_N B m_z + (-1)^{|m_z|+1/2} \frac{e^2qQ}{4} \left(\frac{3\cos^2\theta - 1}{2}\right) \tag{2.22}$$

where $\theta$ is the angle between the magnetic axis and the major axis of the electric field gradient tensor. All the magnetic hyperfine lines are shifted by a quantity

$$|\epsilon| = \frac{e^2qQ}{4} \left(\frac{3\cos^2\theta - 1}{2}\right) \tag{2.23}$$

but the angle $\theta$ and the value of $e^2qQ$ cannot be determined separately from the line positions. A schematic illustration of this is given in Fig.2.4. The presence of a small quadrupole perturbation is easily visible because the spectrum is no longer symmetrical about the centroid. If by chance $\cos\theta = 1/\sqrt{3}$ then the second term in equation (2.22) vanishes and the spectrum appears to be that

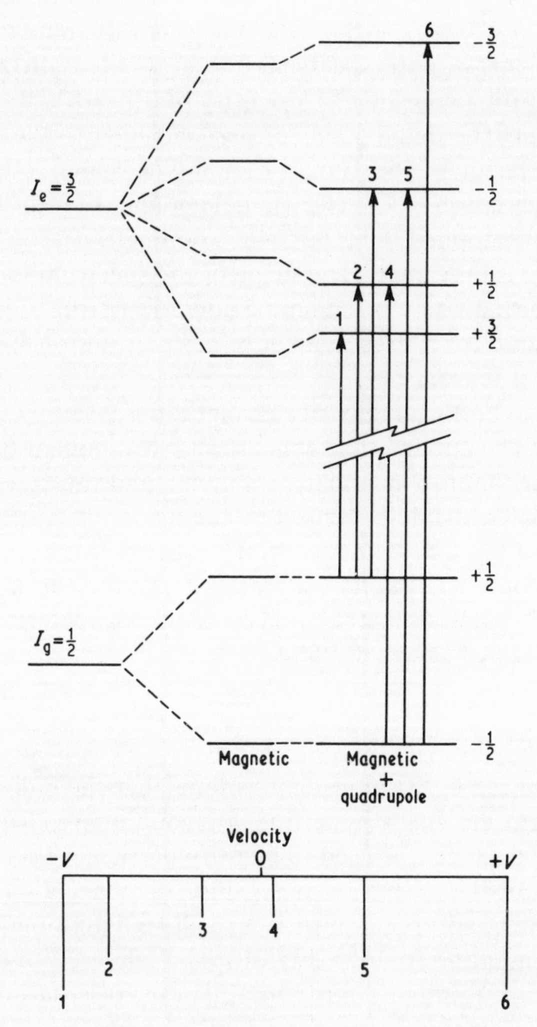


Fig. 2.4 The effect of a small quadrupole perturbation on a $\frac{1}{2} \rightarrow \frac{3}{2}$ magnetic hyperfine splitting. Lines 1 and 6 are shifted to more positive velocity by $+\epsilon$, while lines 2, 3, 4 and 5 are shifted by $-\epsilon$. The difference in the separations 1–2 and 5–6 is therefore $4\epsilon$.

of an unperturbed magnetic hyperfine splitting. Examples of combined magnetic/quadrupole interactions may be found on pp 115–116.

In the preceding section it was found that the sign of $e^2qQ$ and the magnitude of $\eta$ could not be determined from the quadrupole splitting of a $\frac{1}{2} \rightarrow \frac{3}{2}$ transition in a polycrystalline absorber because the spectrum is symmetrical. One means of obtaining this information is to re-measure the spectrum in a large externally applied magnetic field with a magnetic flux density of 30–50 T. It is customary to apply the field parallel or perpendicular to the $\gamma$-ray beam. The resultant spectrum is of complex shape, reflecting the random orientation of the electric field gradient tensor with respect to the applied field, but is not symmetrical, so that the sign of $e^2qQ$ is indicated. The shape of the spectrum may be calculated approximately by summing numerically the individual calculated spectra for a large number of orientations of the electric field gradient tensor.

One of the best examples of this method is given by the $Fe^{57}$ resonance. The applied field splits the $\pm\frac{3}{2}$ component of the quadrupole spectrum into an apparent doublet, and the $\pm\frac{1}{2}$ component into an apparent triplet. Typical calculated spectrum envelopes are shown in Fig.2.5. The spectrum tends towards a symmetrical shape as $\eta$ approaches unity (i.e. when $V_{zz} = -V_{yy}$ the sign of the major axis is indeterminate), and can be used to determine $\eta$ to an accuracy of about ±0.05.

## 2.5 Relative line intensities

It was intimated in the preceding sections that the individual lines in a spectrum showing a hyperfine interaction have characteristic intensities which can be used to identify particular transitions. The magnitudes of the chemical isomer shift, the magnetic hyperfine field and the quadrupole splitting, can all be determined from the line positions alone, and the intensities merely used to confirm the assignment. If it is also desired to obtain the orientation of the magnetic axis or the electric field gradient tensor in an anisotropic sample such as a single crystal, then a more detailed knowledge of line intensities is essential.

The intensity of a particular hyperfine transition between quantized sub-levels is determined by the coupling of the two nuclear angular momentum states [9]. It can be expressed as the product of two terms which are angular-dependent and angular-independent respectively. Since the former averages to unity when all orientations are equally probable, e.g. in a randomly oriented polycrystalline powder sample, it is convenient to consider the angular-independent term first.

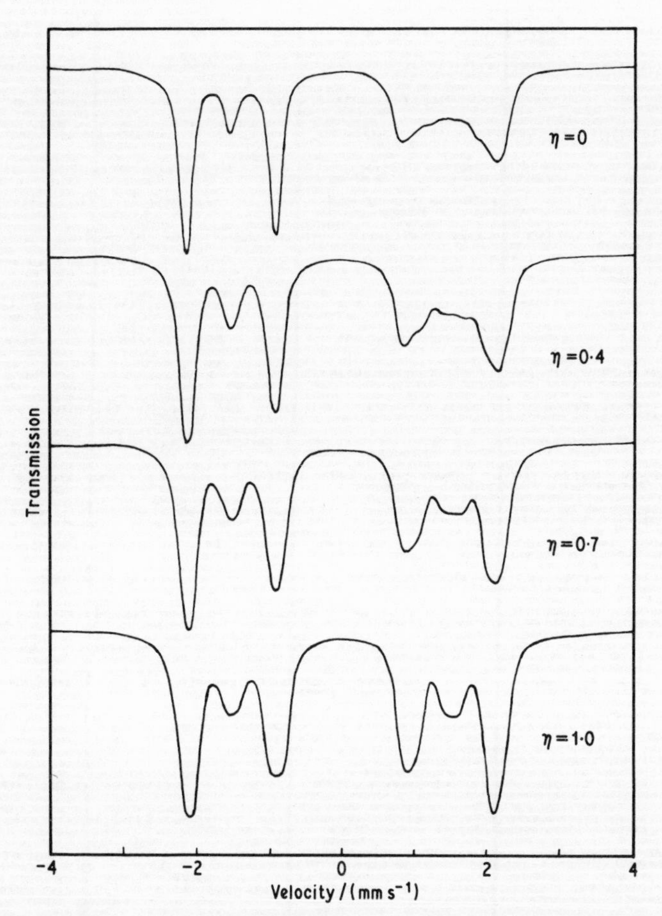


Fig. 2.5 Calculated spectra for a polycrystalline absorber giving an $^{57}Fe$ quadrupole splitting of $\Delta = 3$ mm s$^{-1}$ and in an applied field with a flux density of 5 T parallel to the direction of observation. The spectrum becomes symmetrical as $\eta$ tends towards unity. The sign of $e^2qQ$ is positive in all cases.

The intensity in this instance is given by the square of the appropriate Clebsch-Gordan coefficient [10]

$$\text{Intensity} \propto \langle I_1 J - m_1 m | I_2 m_2 \rangle^2 \qquad (2.24)$$

where the two nuclear spin states $I_1$ and $I_2$ have $I_z$ values of $m_1$ and $m_2$, and their coupling obeys the vector sums $\mathbf{J} = \mathbf{I}_1 + \mathbf{I}_2$ and $m = m_1 - m_2$. $J$ is referred to as the multipolarity of the transition, and the intensity is greater if $J$ is small: if $J = 1$ it is referred to as a dipole transition, while with $J = 2$ it is a quadrupole transition. Most of the Mössbauer transitions take place without a change in parity, so that the radiation is classified as a magnetic dipole (M1) or electric quadrupole (E2) transition. One of the few exceptions is the 25.65-keV $\frac{5}{2}+ \rightarrow \frac{5}{2}-$ transition in $^{161}Dy$ which is an electric dipole (E1) transition. In some instances such as $^{99}Ru$(90-keV) and $^{197}Au$(77-keV) the radiation is a mixture of M1 and E2, and both terms must be added in the appropriate proportion. The selection rule for an M1 or E1 transition is $\Delta m_z = 0, \pm 1$, and for an E2 transition is $\Delta m_z = 0, \pm 1, \pm 2$.

The most frequently used coefficients are those for the $\frac{1}{2} \rightarrow \frac{3}{2}$ M1 transition, and these are given in Table 2.2 (coefficients for other spin states have been tabulated in [11]). $I_1$ may be either the ground or excited state spin. Although there are nominally eight transitions, the $+\frac{3}{2} \rightarrow -\frac{1}{2}$ and $-\frac{3}{2} \rightarrow +\frac{1}{2}$ transitions have a zero probability (forbidden). The six finite coefficients, $C^2$, which express the angular-independent intensity have a total probability of unit intensity and give directly the 3:2:1:1:2:3 intensity ratios for a magnetic hyperfine splitting, shown in Fig.2.2. The corresponding terms for a quadrupole spectrum are obtained by summation and give a 1:1 ratio.

The angular dependent terms, $\Theta(J, m)$, are expressed as the radiation probability in a direction at an angle $\theta$ to the quantization axis (i.e. the magnetic field axis or $V_{zz}$: note that the values in the table in the latter case are only correct if $\eta = 0$). The intensity for a polycrystalline sample is obtained by integration over all $\theta$ to obtain $\overline{\Theta(J, m)}$; e.g.

$$\tfrac{3}{2}\,\overline{\sin^2\theta} = \frac{1}{4\pi}\int_0^{2\pi}\int_0^{\pi} (\tfrac{3}{2}\sin^2\theta)\sin\theta\,d\theta\,d\phi = 1 \tag{2.25}$$

and the total of emitted radiation is independent of $\theta$ and normalized to unity, i.e.

$$\sum_{m_1 m_2} \tfrac{1}{4}\langle I_1 J - m_1 m | I_2 m_2\rangle^2 \overline{\Theta(J, m)} = 1 \tag{2.26}$$

Coefficients such as those in Table 2.2 are necessary to interpret the angular dependence of the spectrum from a single crystal or oriented absorber. For example, a magnetically ordered metal alloy or oxide absorber may often be 'polarized' by magnetizing in a small external magnetic field to give a unique direction of the internal field. The expected line intensities can then be predicted from Table 2.2 to be in the ratios $3{:}x{:}1{:}1{:}x{:}3$ where $x = 4\sin^2\theta/(1 + \cos^2\theta)$; in particular the $m = 0$ transitions have a zero intensity when observed along the direction of the field ($\theta = 0°$) and a maximum intensity perpendicular to the field ($\theta = 90°$). This is illustrated schematically in Fig. 2.6.

The equivalent behaviour in the quadrupole spectrum is a 1:3 ratio for the γ-ray axis parallel to the direction of $V_{zz}$ and a 5:3 ratio perpendicular to $V_{zz}$. In this instance the angular dependence

Table 2.2 The relative probabilities for a $\frac{1}{2}, \frac{3}{2}$ transition

Magnetic spectra (M1)

| $m_2$ | $-m_1$ | $m$ | $C$ (1) | $C^2$ (2) | $\Theta(J, m)$ (2) |
|---|---|---|---|---|---|
| $+\frac{3}{2}$ | $+\frac{1}{2}$ | +1 | 1 | $\frac{1}{4}$ | $\frac{3}{4}(1+\cos^2\theta)$ |
| $+\frac{1}{2}$ | $+\frac{1}{2}$ | 0 | $\sqrt{\frac{2}{3}}$ | $\frac{1}{6}$ | $\frac{3}{2}\sin^2\theta$ |
| $-\frac{1}{2}$ | $+\frac{1}{2}$ | −1 | $\sqrt{\frac{1}{3}}$ | $\frac{1}{12}$ | $\frac{3}{4}(1+\cos^2\theta)$ |
| $-\frac{3}{2}$ | $+\frac{1}{2}$ | −2 | 0 | 0 | — |
| $+\frac{3}{2}$ | $-\frac{1}{2}$ | +2 | 0 | 0 | — |
| $+\frac{1}{2}$ | $-\frac{1}{2}$ | +1 | $\sqrt{\frac{1}{3}}$ | $\frac{1}{12}$ | $\frac{3}{4}(1+\cos^2\theta)$ |
| $-\frac{1}{2}$ | $-\frac{1}{2}$ | 0 | $\sqrt{\frac{2}{3}}$ | $\frac{1}{6}$ | $\frac{3}{2}\sin^2\theta$ |
| $-\frac{3}{2}$ | $-\frac{1}{2}$ | −1 | 1 | $\frac{1}{4}$ | $\frac{3}{4}(1+\cos^2\theta)$ |

Quadrupole spectra (M1) when $\eta = 0$

| Transition | $C^2$ (2) | $\Theta(J, m)$ (2) |
|---|---|---|
| $\pm\frac{1}{2}, \pm\frac{1}{2}$ | $\frac{1}{2}$ | $\frac{1}{2} + \frac{3}{4}\sin^2\theta$ |
| $\pm\frac{3}{2}, \pm\frac{1}{2}$ | $\frac{1}{2}$ | $\frac{3}{4}(1+\cos^2\theta)$ |

(1) The Clebsch-Gordan coefficient $\langle\frac{1}{2}1 - m_1 m|\frac{3}{2}m_2\rangle$
(2) $C^2$ and $\Theta(J, m)$ are the angular-independent and angular-dependent terms normalized to a total radiation probability of

$$\sum_{m_1 m_2} C^2\Theta(J, m) = 1$$

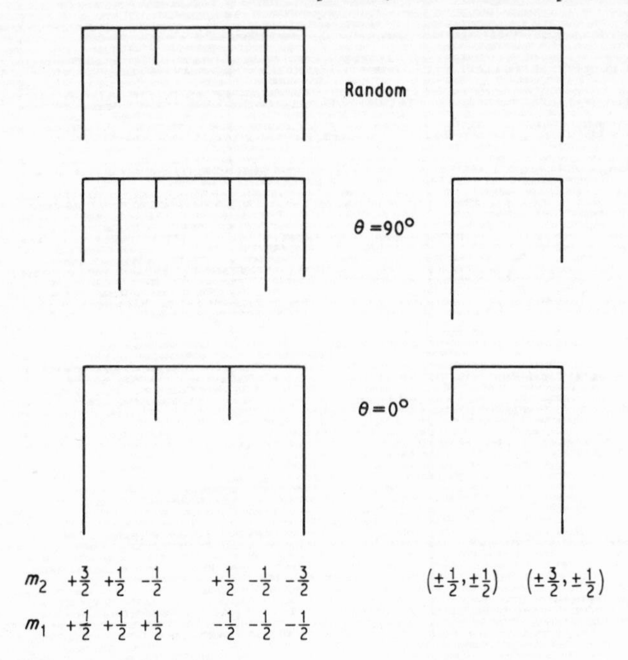


Fig. 2.6 The effect of orientation upon the relative line intensities of a magnetic hyperfine splitting and a quadrupole splitting of a $\frac{3}{2} \rightarrow \frac{1}{2}$ transition in an oriented absorber with a unique principal axis sytem.

can be used to determine which of the two lines is the $\pm\frac{1}{2} \rightarrow \pm\frac{3}{2}$ transition, and hence to obtain the sign of $e^2qQ$.

If there are combined magnetic and quadrupole interactions, or the asymmetry parameter $\eta$ is non-zero, the previous discussion is no longer strictly valid. In these circumstances the energy states are no longer 'pure' states with a defined quantum number, and must be represented as linear combinations of terms. Likewise the intensity is obtained by an amplitude summation to allow for interference effects. These calculations are somewhat complex [11] and will not be described here.

The total cross-section for resonant absorption is a nuclear constant, so that the individual hyperfine lines are proportionately weaker than in an unsplit resonance line. This unfortunately necessitates a longer counting time to resolve the extra detail. In this discussion the absorption intensity has been considered for a single nucleus, but it is also important to consider the effects of finite absorber thickness (see p.11–12). The effective thickness of the absorber $T_a = n_a \sigma_0 f_a$ is now multiplied by the fractional line intensity for each component $C^2\Theta(J, m)$, with the result that saturation of the absorption intensity occurs more quickly for the more intense resonance lines. This is seen as a dependence of the relative line intensities on increasing thickness which tends to decrease any asymmetry. In addition, the more intense lines have a larger linewidth due to the thickness broadening effect. Because of this, a quadrupole doublet spectrum of for example a single crystal with the direction of observation parallel to $V_{zz}$ will show a line intensity ratio much less than 1:3.

An additional complication which also affects the intensities appreciably is polarization of the incident γ-ray beam as it progresses through the crystal [12]. For these reasons, unless it is proposed to attempt a somewhat difficult thickness correction, it is necessary to obtain relative intensity data from very thin absorbers. However, the possibility of underestimation of any asymmetry should always be borne in mind.

There are two other important influences on relative line intensities which are of particular importance in powders. If the compacted polycrystalline powder has a tendency towards partial orientation (texture), then the values of $\Theta(J, m)$ do not average to unity, and there is some residual asymmetry in the spectrum which is angular dependent. This effect can be quite large, and exceedingly difficult to eliminate in fibrous or platelike materials. A second effect which is similar in many ways is the Goldanskii-Karyagin effect, in which anisotropy of the recoilless fraction results in an asymmetry which is not angular dependent but is usually small. This is discussed in more detail in Chapter 6. In many instances deviations from the 1:1 line intensity ratio in a $\frac{1}{2} \rightarrow \frac{3}{2}$ quadrupole doublet have been attributed to the Goldanskii-Karyagin effect without convincing proof that there was no residual texture in the sample.

## References

[1] Pound, R. V. and Rebka, G. A. (1959) *Phys. Rev. Letters*, **3**, 554.
[2] Kistner, O. C. and Sunyar, A. W. (1960) *Phys. Rev. Letters*, **4**, 412.
[3] Breit, G. (1958) *Rev. Mod. Phys.*, **30**, 507.
[4] Reddy, K. R., Barros, F. de S. and DeBenedetti, S. (1966) *Phys. Letters*, **20**, 297.
[5] Pound, R. V. and Rebka, G. A. (1960) *Phys. Rev. Letters*, **4**, 274.

[6] Abragam, A. (1961) *The Principles of Nuclear Magnetism.* Clarendon Press, Oxford.
[7] Kubo, M. and Nakamura, D. (1966) *Adv. Inorg. Chem. and Radiochem.*, 8, 257.
[8] Ingallis, R. (1962) *Phys. Rev.*, **128**, 1155.
[9] Rose, M. E. (1955) *Multipole Fields.* Chapman and Hall, London.
[10] Condon, E. U. and Shortley, G. H. (1935) *The Theory of Atomic Spectra.* Cambridge Univ. Press.
[11] Greenwood, N. N. and Gibb, T. C. (1971) *Mössbauer Spectroscopy*, Chapman and Hall, London.
[12] Housley, R. M., Grant, R. W. and Gonser, U. (1969) *Phys. Rev.*, **178**, 514.

CHAPTER THREE

# *Molecular Structure*

In the preceding chapter it was shown how the hyperfine interactions in a Mössbauer spectrum contain information about the immediate electronic environment of the resonant nucleus. In particular, they can be related to the geometrical symmetry and to the nature of the chemical bonding involved. For example, a change in the formal oxidation state of the resonant nucleus (removal or addition of a valence electron) may be expected to result in a change in the *s*-electron density at the nucleus and thence in the chemical isomer shift. The remaining chapters of this book are therefore concerned with various aspects of the chemical application of Mössbauer spectroscopy. In this chapter, some examples of its use to determine molecular structure, i.e. the geometrical disposition of atoms in a compound, are discussed. The study of electronic structure and the way in which these atoms are chemically bonded is deferred until the following chapter.

The successful determination of structural data from the Mössbauer spectrum of a molecular complex is very much governed by individual circumstance. In the majority of compounds studied there has been only one suitable resonant isotope. The advantage of measuring parameters specific to particular atoms is offset by the difficulty of determining the number and disposition of the atoms or groups to which they are bonded. This is particularly true in diamagnetic compounds where the primary parameters of the chemical isomer shift and the quadrupole splitting are a function of all the valence electrons, which are of course usually participating in covalent bonding to more than one atom. The unpaired electrons in the valence shells of paramagnetic compounds produce characteristic effects which are discussed in Chapter 5.

The resonant atom in a diamagnetic compound gives a simple spectrum, usually with a quadrupole splitting. In the event that there are several non-equivalent resonant atoms in the structure, then each should give a different contribution to the spectrum; indeed the observation of two quadrupole spectra from a compound with two resonant atoms is proof that these are not in the same environment. The converse, that one quadrupole spectrum indicates identity of environment, is not necessarily true, since any differences may be less than the resolution of the measurement.

## 3.1 Iron carbonyls and derivatives

The 14.4-keV $^{57}Fe$ resonance has been widely used in structural studies of the iron carbonyls and their derivatives where multiple iron environments are quite common. A good example is provided [1] by the spectra of polycrystalline $[Fe(py)_6]^{2+}[Fe_4(CO)_{13}]^{2-}$ and $[NEt_4]_2^+[Fe_4(CO)_{13}]^{2-}$ illustrated in Fig.3.1. Quadrupole splitting of the $I_g = \frac{1}{2} \rightarrow I_e = \frac{3}{2}$ transition results in a symmetrical doublet for each distinct iron environment. The extra pair of lines in the former case are thus a quadrupole doublet from the high-spin $Fe^{2+}$ complex cation. This assignment is borne out both by comparing the value of the chemical isomer shift with data from other $Fe^{2+}$ compounds, and by the approximate absorption area which is one quarter of that due to the anion. However, the anion itself is known to be a triangular-based pyramid of $Fe(CO)_3$ groups with the thirteenth carbonyl in a triply-bridging configuration below the basal plane. Although there are three equivalent iron atoms and a unique one, it is not possible to visually resolve the apparently small differences in the Mössbauer spectrum.

As a further example, the hydride anion $[Fe_2(CO)_8H]^-$ shows only a simple quadrupole splitting in the $^{57}Fe$ spectrum, suggesting that there is only a single iron environment [2]. The three most likely structures are (1)–(3). Structures (1) and (2) are derived from that of $Fe_2(CO)_9$ by substitution of a bridging and a terminal carbonyl group respectively. Only (1) and (3) have identical iron environments, but (2) cannot be completely eliminated as simply even though data for other carbonyl hydrides leads one to predict

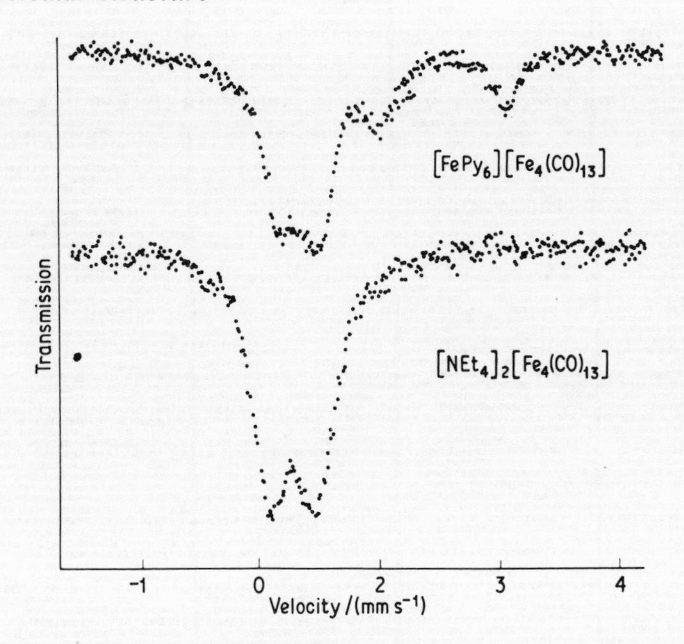


Fig. 3.1 $^{57}$Fe spectra at 80 K for the $[Fe(py)_6]^{2+}$ and $[NEt_4]^+$ salts of $[Fe_4(CO)_{13}]^{2-}$. (py = pyridine) Note the extra lines showing quadrupole splitting from the cation in the former, and the failure to distinguish the unique iron atom in the carbonyl anion. ([1], Fig. 2.10).

1

2

3

a detectable difference in parameters. In a situation like this it is often advantageous to consider the Mössbauer data in combination with other spectroscopic measurements. In the event, the presence of bridging carbonyl bands in the infra-red spectrum eliminates structure (3), and the total number of bands corresponds to prediction from the symmetry of structure (1). As may be seen from Table 3.1, the Mössbauer parameters for $[Fe_2(CO)_8H]^-$ are similar to those for $Fe_2(CO)_9$. The replacement of a bridging carbonyl in $Fe_2(CO)_9$ by hydrogen causes a reduction in chemical isomer shift, $\delta$, and an increase in the quadrupole splitting, $\Delta$, exactly analagous to the observed changes in $Fe_3(CO)_{12}$ and $[Fe_3(CO)_{11}H]^-$. As a result, the Mössbauer data in conjunction with the infra-red evidence constitute satisfactory proof that structure (1) is correct.

The chemical isomer shift is not a direct function of stereochemistry because it involves the electron wavefunctions with spherical symmetry. Nevertheless, a significant difference in shift between two related compounds is usually evidence of either a change in co-ordination number, or in the co-ordinated ligands, or both. Such a change may have a large effect on the *p*- and *d*-electron shielding or on the bond hybridization. On the other hand, the quadrupole splitting derives from asymmetric occupation of *p*-, *d*- and *f*-electron orbitals and is therefore related directly to the geometry of the compound. A study of these parameters for a large number of related compounds reveals systematic relationships

Table 3.1 Mössbauer parameters from the 14.4-keV $^{57}Fe$ resonance at 80 K in some derivatives of $Fe_2(CO)_9$ and $Fe_3(CO)_{12}$

| | Outer doublet | | Inner doublet | |
|---|---|---|---|---|
| Compound | Δ* | δ* | Δ* | δ* |
| $Fe_2(CO)_9$ | 0.42 | +0.16 | – | – |
| $[Fe_2(CO)_8H]^-$ | 0.50 | +0.06 | – | – |
| $Fe_3(CO)_{12}$ | 1.13 | +0.11 | ~0 | +0.05 |
| $[Fe_3(CO)_{11}H]^-$ | 1.41 | +0.04 | 0.16 | +0.02 |
| $[(OC)_3FePMe_2Ph]_3$ | 1.15 | +0.09 | 0.57 | +0.02 |
| $(ffars)Fe_3(CO)_{10}$ | 1.52 | +0.16 | ~0 | +0.03 |
| $HFe_3(CO)_{10}CNMe_2$ | 0.94 | –0.04 | 0.16 | +0.04 |

* Values in mm $s^{-1}$. δ is given relative to iron metal at room temperature. ffars = 1,2-bis(dimethylarsino)tetrafluorocyclobutene.

with such features as co-ordination number, site symmetry, and the nature of the bonding groups. Diagnostic interpretation of the spectra from new uncharacterized compounds can then become possible.

As an example, the Mössbauer parameters for iron carbonyls are not very sensitive to substitution, but the quadrupole splitting is often indicative of geometry [3]. An iron atom in 6-coordination tends to give only a small quadrupole splitting as seen in for example $Fe_2(CO)_9$, $[Fe_2(CO)_8H]^-$ and $(OC)_3Fe(PMe_2)_2Fe(CO)_3$. The latter (structure 4) has an Fe-Fe bond to complete the 6-co-ordination (with a quadrupole splitting of $\Delta = 0.68$ mm $s^{-1}$). In

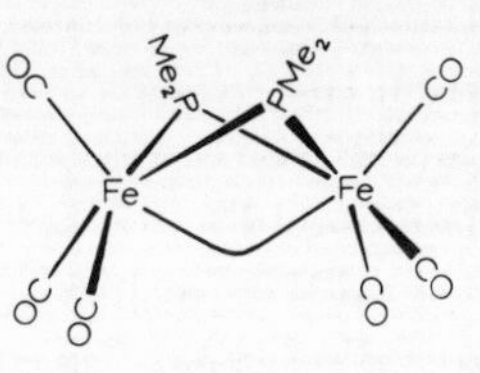

4

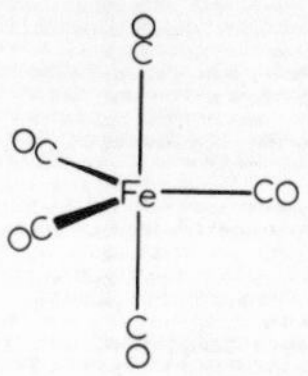

5

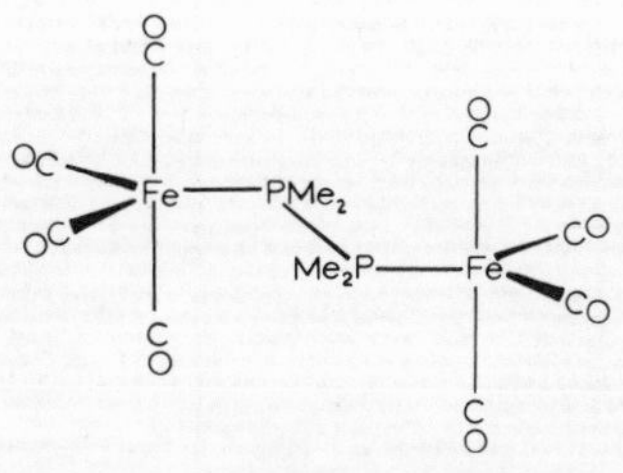

6

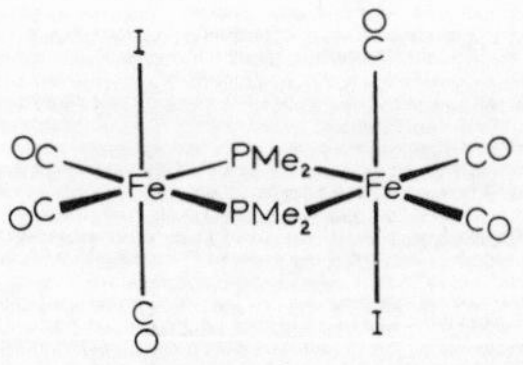

7

contrast, $Fe(CO)_5$ ($\Delta = 2.57$ mm $s^{-1}$) and $(OC)_4Fe.PMe_2.PMe_2.Fe(CO)_4$ ($\Delta = 2.58$ mm $s^{-1}$) are 5-coordinated, but $(OC)_3IFe(PMe_2)_2FeI(CO)_3$ ($\Delta = 0.99$ mm $s^{-1}$) is again 6-coordinated with a small splitting (structures 5–7). However, the range of structural types is so large that reliable generalizations by empiricism can become difficult.

In appropriate instances it is helpful to determine the sign of $e^2qQ$ and the magnitude of the asymmetry parameter, $\eta$, by applying a large external magnetic field (see p.38). An example is given by the results in Table 3.2 for some triphenylphosphine and

Table 3.2 $^{57}Fe$ quadrupole coupling parameters for phosphine derivatives of $Fe(CO)_5$

| Compound | $\frac{1}{2}e^2qQ/(\mathrm{mm\ s^{-1}})$ | $\eta$ |
|---|---|---|
| $Fe(CO)_5$ | +2.60 | 0.0 |
| $Ph_3PFe(CO)_4$ | +2.54 | 0.0 |
| $(Ph_3P)_2Fe(CO)_3$ | +2.76 | 0.0 |
| $(diphos)[Fe(CO)_4]_2$ | +2.46 | 0.0 |
| $(diphos)Fe(CO)_3$ | −2.12 | 0.8 |

1,2-bisdiphenylphosphinoethane (diphos) substituted derivatives of $Fe(CO)_5$ [4]. The magnitude of the quadrupole splitting is insensitive to the substitution, but when the axial symmetry of the trigonal bipyramid is broken by a chelated substitution as in $(diphos)Fe(CO)_3$ the sign of $e^2qQ$ changes and $\eta$ increases dramatically. However, it is still not possible to distinguish whether the bi-dentate substitution is axial-equatorial or equatorial-equatorial. Nevertheless, it is clear that the triphenylphosphine groups in $(Ph_3P)_2Fe(CO)_3$ are both in the axial positions.

The usefulness of Mössbauer spectroscopy in following structural changes in a family of related compounds can be well illustrated by some studies made on derivatives of the iron carbonyl, $Fe_3(CO)_{12}$. The parent carbonyl shows a spectrum with three lines of equal intensity [5]. This is illustrated in Fig.3.2. Comparison with other carbonyl data allows confident assignment of the outer lines to a quadrupole splitting of two apparently identical iron atoms, and the centre line to the third atom in a more symmetrical environment such that any quadrupole splitting is not resolvable. These observations were considerably at variance with the then accepted crystal structure (which was incomplete due to complications from crystal disorder effects) [7]. This structure proposed an equilateral triangle of iron atoms, each with two terminal carbonyls and two carbonyls bridging to each of the other iron atoms, so that all three iron atoms were equivalent.

The spectrum of the related $[Fe_3(CO)_{11}H]^-$ anion is similar to that of $Fe_3(CO)_{12}$ but with a larger splitting in the outer doublet [8]. The parameters are given in Table 3.1 for comparison. It can be deduced that the hydrogen has replaced a carbonyl group bridging to the two equivalent iron atoms, and the structure pro-

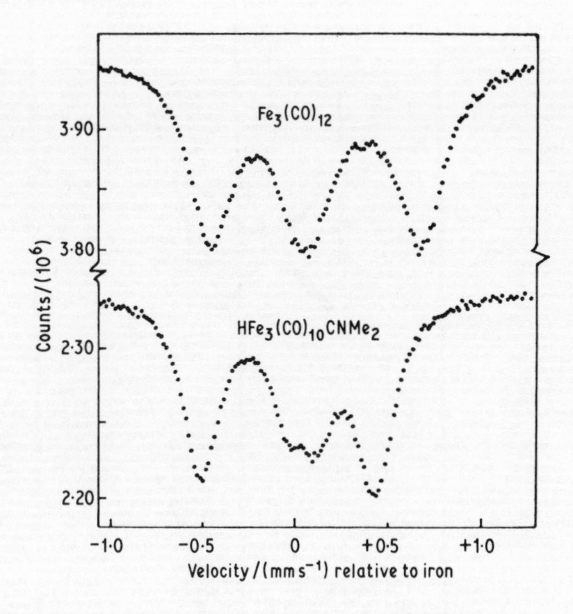


Fig. 3.2 $^{57}$Fe Mössbauer spectra at 80 K for $Fe_3(CO)_{12}$ and $HFe_3(CO)_{10}CNMe_2$. The two compounds are clearly related structurally. The central line in $Fe_3(CO)_{12}$ does not show a resolved quadrupole splitting. ([6], Fig. 1).

posed for the hydride (structure 8) has been verified by X-ray techniques. A recent re-investigation of the parent $Fe_3(CO)_{12}$ has confirmed [9] the now suspected structure (9), showing the isosceles triangular arrangement and the close relationship to $[Fe_3(CO)_{11}H]^-$.

The substituted derivative $Fe_3(CO)_9(PMe_2Ph)_3$ can have one of many possible structures depending on where the substitution takes place. The spectrum is again closely similar to that of $Fe_3(CO)_{12}$, but shows a small decrease in the chemical isomer shift at both types of iron atom, as well as small changes in quadrupole splitting (see Table 3.1). Here again, the tentative structure (10) involving substitution at each iron atom has been fully confirmed at a later date by X-ray analysis [10].

Another derivative, (ffars)$Fe_3(CO)_{10}$ is also clearly related to $Fe_3(CO)_{12}$ [11]. The two equivalent iron atoms remain so, but with a considerable increase in quadrupole splitting, leading to the proposed structure (11).

A final example is given by a compound initially reported as

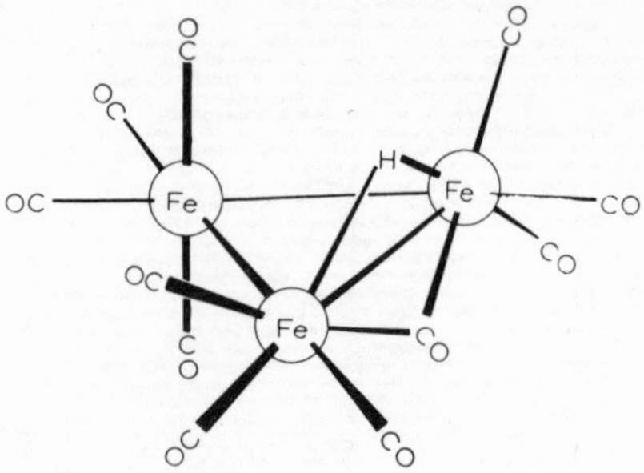

8

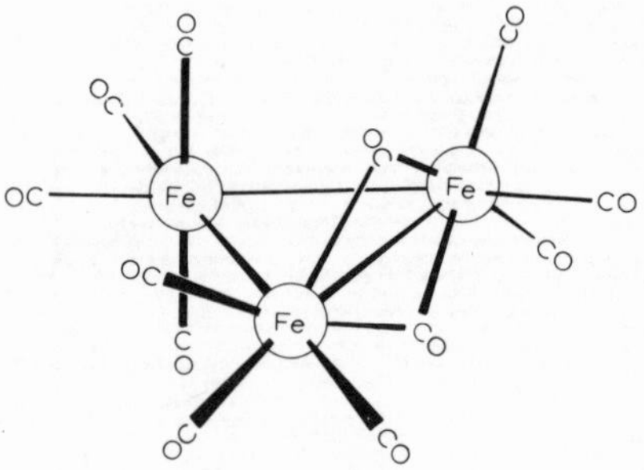

9

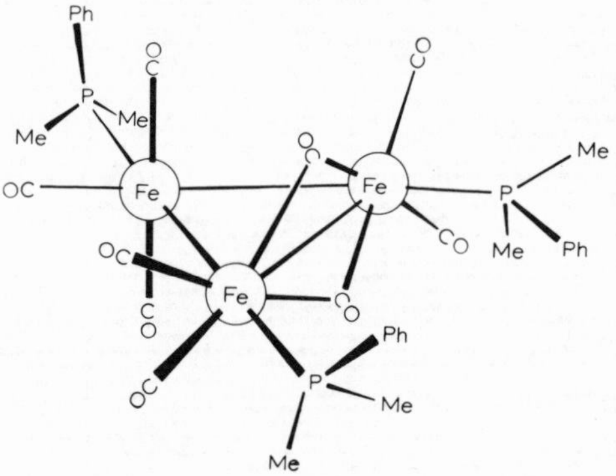

10

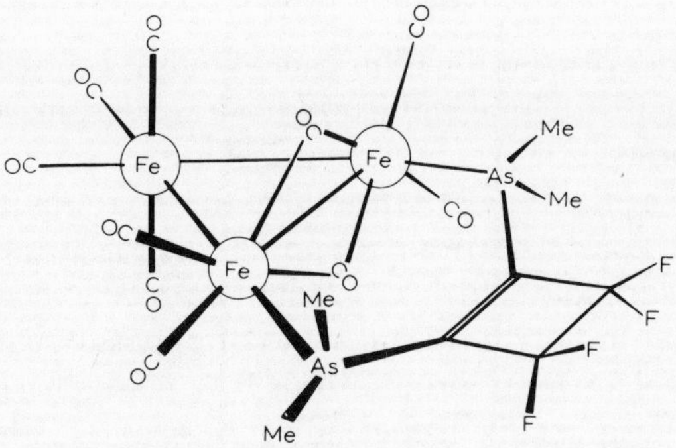

11

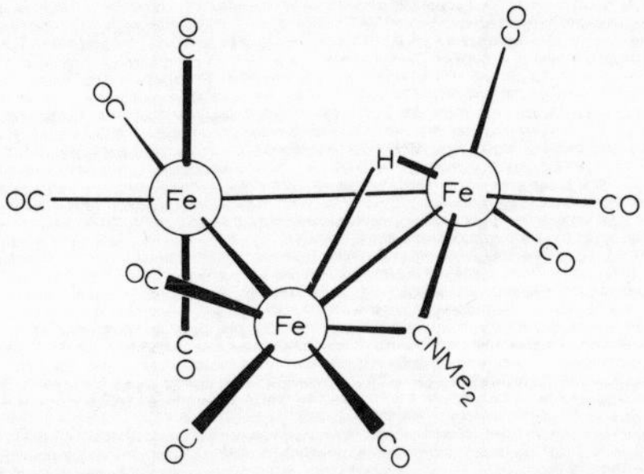

12

$HFe_3(CO)_{11}NMe_2$. The suggested structure featured a terminal $-NMe_2$ group. The Mössbauer spectrum indicates a strong similarity to $Fe_3(CO)_{12}$ with any structural change affecting the bridged iron atoms equally (see Fig.3.2), and is incompatible with the initial proposition. Further investigation by mass-spectrometry has established that the compound is in fact $HFe_3(CO)_{10}CNMe_2$ [5], and combination of the Mössbauer data with the fragmentation pattern from the mass spectrometer led to the novel structure (12). The mass spectrum shows stepwise loss of the carbonyl groups to give a skeletal fragment $HFe_3CNMe_2^+$, and other positively identified fragments were $HFe_3C^+$ and $HFe_2(CO)CNMe_2^+$, good evidence for the adopted formulation.

## 3.2 Geometrical isomerism in Fe and Sn compounds

In many series of co-ordination compounds it is possible to find examples of a structural ambiguity from isomerism. One of the common forms is *cis-/trans-*geometrical isomerism in octahedral compounds of the type $MA_2B_4$ where the A groups are either adjacent (*cis*) or diametrically opposed (*trans*). It is easy to show that the quadrupole splitting of atom M can be expected to be twice as great in the *trans*-isomer and of opposite sign to the *cis*-isomer. An elementary proof of this by a point charge calculation is given in the appendix at the end of the chapter, where the principal values of the electric field gradient tensors in $MAB_5$, *cis*-$MA_2B_4$ and *trans*-$MA_2B_4$ are found to be in the ratio 1:–1:2 respectively. The asymmetry parameter is zero in all three cases.

Experimental verification of this was first given by Berrett and Fitzsimmons [12] for some isocyanide complexes of iron:

| | |
|---|---|
| $[Fe(CN)(CNEt)_5](ClO_4)$ | $\Delta = 0.17$ mm s$^{-1}$ |
| *cis*-$Fe(CN)_2(CNEt)_4$ | $\Delta = 0.29$ mm s$^{-1}$ |
| *trans*-$Fe(CN)_2(CNEt)_4$ | $\Delta = 0.56$ mm s$^{-1}$ |

and in one instance the difference in sign between the *cis*- and *trans*-isomers has been verified by the applied field method [13]:

| | |
|---|---|
| $[FeCl(p\text{-}MeO.C_6H_4.NC)_5](ClO_4)$ | $\Delta = 0.70$ mm s$^{-1}$ |
| *cis*-$FeCl_2(p\text{-}MeO.C_6H_4.NC)_4$ | $\Delta = -0.83$ mm s$^{-1}$ |
| *trans*-$FeCl_2(p\text{-}MeO.C_6H_4.NC)_4$ | $\Delta = +1.59$ mm s$^{-1}$ |

This general rule for distinguishing isomers will remain valid provided that the geometry is close to octahedral and provided that the electron distribution in one M-A bond is not significantly influenced by its geometrical relationship to the other, i.e. there is no large '*trans*-effect' (note that a gross distortion of the geometry may lead to a large value of the asymmetry parameter and/or a change in the principal axis system of the electric field gradient tensor).

Similar results have been found in 6-coordinate organotin compounds using the 23.87-keV $I_g = \frac{1}{2} \rightarrow I_e = \frac{3}{2}$ resonance of $^{119}Sn$. An empirical study of observed quadrupole splittings reveals a characteristic distinction between on the one hand groups bonding through carbon such as alkyl and aryl groups, and on the other hand halides and nitrogen and oxygen bonded donor ligands. In consequence, a compound such as $Ph_2SnCl_2$.phen behaves as if it were of the type $MA_2B_4$ where *both* the chlorine atoms and the 1,10-phenanthroline are classed as B ligands. This rather peculiar effect will be mentioned again later in the chapter. Such complexes also show the 1:2 ratio for *cis*- and *trans*-structures as may be seen from Table 3.3, and it is therefore possible to derive unknown stereochemistries [14]. The earlier assignment of *trans*-phenyl groups to $Ph_2Sn(acac)_2$ is evidently incorrect. The *cis*-isomers in chelated tin compounds are usually distorted from an octahedral geometry. As described on p.67 this may result in a change in the sign of the quadrupole splitting so that it corresponds to that of a *trans*-isomer; however the magnitude of the splitting still remains approximately the same as expected for a more regular geometry, and the method is still diagnostic of configuration if this problem is borne in mind.

As a cautionary note, mention must be made of the *syn*- and *anti*-isomers of the iron cyclopentadienyl derivative Cp(OC)Fe-$(PMe_2)_2$Fe(CO)Cp (structure 13) which have the cyclopentadienyl ring on the same and opposite sides of the $Fe_2P_2$ plane respectively. In this case the $^{57}Fe$ spectrum is the same in both compounds. The immediate Fe environment is identical, and is not influenced significantly by geometrical changes at a much greater distance [3].

Table 3.3 The relationship between the quadrupole splitting and *cis-/trans*-isomerism in organotin compounds

| Compound | $\Delta/(mm\ s^{-1})$ | Stereochemistry by Mössbauer | Stereochemistry by other methods |
|---|---|---|---|
| $Ph_2Sn(acac)_2$ | 2.14 | *cis* | *trans* |
| $Me_2Sn(acac)_2$ | 3.93 | *trans* | *trans* |
| $Ph_2SnCl_2$.phen | 3.70 | *trans* | |
| $Ph_2SnCl_2$.bipy | 3.90 | *trans* | |
| $Me_2SnCl_2$.phen | 4.03 | *trans* | |
| $Pr^n_2Sn.oxin_2$ | 2.20 | *cis* | |
| $Ph_2Sn.oxin_2$ | 1.78 | *cis* | *cis* |
| $Bu^n_2Sn.oxin_2$ | 2.21 | *cis* | |
| $Me_2Sn.oxin_2$ | 1.98 | *cis* | *cis* |

acac = acetylacetone, phen = 1,10-phenanthroline, bipy = 2,2′-bipyridyl, oxin = 8-hydroxyquinoline.

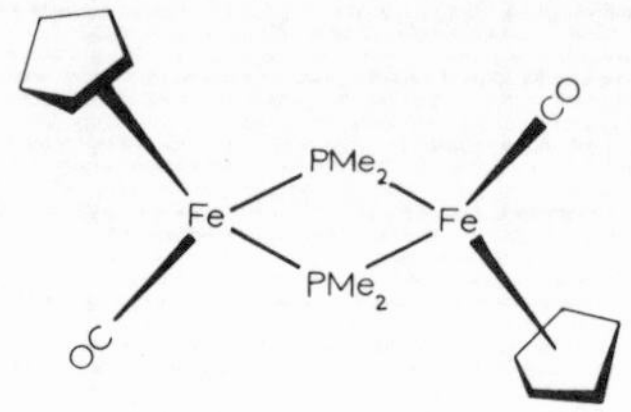

13

### 3.3 Linkage isomerism in cyano-complexes of Fe

Some simple inorganic ligands such as $CN^-$ and $NCS^-$ are capable of co-ordinating to a metal cation from either end of the molecule, and this leads to a possible structural ambiguity. The hexacyanoferrate(II) complexes with the transition metals usually form a polymeric lattice structure with $Fe^{II}(CN)_6^{4-}$ anions co-ordinated through the nitrogens to other metal cations. Excess cations required for charge neutrality can be accommodated in the large interstitial sites. The idealized cubic structure is shown in Fig.3.3. There exists the possibility of a kind of isomerism known as linkage isomerism involving the interchange of the cations between the sites co-ordinated to carbon and to nitrogen. This is equivalent to reversing the C-N molecule, thereby changing the metal co-ordination from $[M(CN)_6]^{n-}$ to $[M(NC)_6]^{n-}$.

The four possible electronic configurations of iron in these complex cyanides are low-spin iron(II) or iron(III) co-ordinated to carbon, and high-spin $Fe^{2+}$ and $Fe^{3+}$ co-ordinated to nitrogen or in interstitial sites. High-spin $Fe^{2+}$ and $Fe^{3+}$ have characteristically different chemical isomer shifts and are easily distinguished, but empirical observations on simple hexacyanoferrate(II) and hexacyanoferrate(III) compounds show that these low-spin configurations will have similar chemical isomer shifts. However, a hexacyanoferrate(II) usually gives a single-line resonance, whereas the hexacyanoferrate(III) may show a substantial quadrupole splitting. It is therefore possible to determine which sites are occupied by iron, and also the cation oxidation states, in a compound with the structure shown in Fig.3.3.

For example, the reaction product precipitated from solutions of $TiCl_3$ and $H_3Fe(CN)_6$ gives the single sharp absorption corresponding to a hexacyanoferrate(II) [15], and is clearly $Ti^{4+}[Fe^{II}(CN)_6]^{4-}$

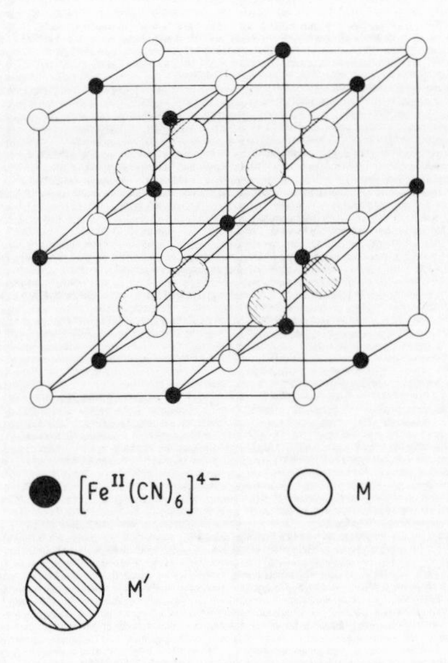


Fig. 3.3 The crystal structure of the transition-metal hexacyanoferrates.

A mutual redox reaction has taken place. The parameters are given in Table 3.4. $TiCl_3$ or $TiCl_4$ with $H_4Fe(CN)_6$ give the identical product, showing that oxidation occurs with the former. However, $CoCl_2$ with $K_4Fe(CN)_6$ and $K_3Fe(CN)_6$ gives a hexacyanoferrate(II) and a hexacyanoferrate(III) respectively, $Co^{2+}_2[Fe^{II}(CN)_6]^{4-}$ and $Co^{2+}_3[Fe^{III}(CN)_6]^{3-}_2$, and no redox reaction occurs.

An interesting technique which has been successfully applied to these systems is that of selective isotopic enrichment. The reaction of $Fe_2(SO_4)_3$ with $K_3Fe(CN)_6$ gives a hexacyanoferrate(III) complex, $Fe^{3+}[Fe^{III}(CN)_6]^{3-}$. The spectrum (Fig.3.4) has a complex shape because components from the two types of iron overlap. The two resolved peaks are due to a quadrupole splitting of the $Fe^{3+}$ ion. The low-spin component is not resolved, and it is not easy to distinguish the oxidation state. However, if the preparation is repeated using $K_3Fe(CN)_6$ made with $^{57}Fe$ enriched iron, the absorption cross-section for the $[Fe^{III}(CN)_6]^{3-}$ can be increased by a factor of about 40. The observed resonance spectrum is then completely dominated by the low-spin component to the effective exclusion of the $Fe^{3+}$ lines, and as shown in Fig.3.4 the quadrupole splitting is then clearly seen, thereby confirming the presence of $[Fe^{III}(CN)_6]^{3-}$.

Table 3.4 Mössbauer parameters* for complex iron cyanides

| Compound | $Fe^{II}$ Δ | $Fe^{II}$ δ | $Fe^{III}$ Δ | $Fe^{III}$ δ | $Fe^{3+}$ Δ | $Fe^{3+}$ δ |
|---|---|---|---|---|---|---|
| $Ti[Fe(CN)_6]$ | 0 | −0.01 | | | | |
| $Co_2[Fe(CN)_6]$ | ~0 | −0.01 | | | | |
| $Co_3[Fe(CN)_6]_2$ | | | 0.85 | −0.08 | | |
| $Fe[Fe(CN)_6]$ | | | 0.43 | −0.06 | 0.52 | +0.50 |
| Prussian blue | 0 | −0.08 | | | 0.57 | +0.49 |
| Turnbull's blue | 0 | −0.07 | | | 0.51 | +0.49 |

* All data at 77 K in mm $s^{-1}$ δ is relative to iron metal at room temperature.

The classic compounds 'Prussian blue' made from $Fe^{3+}$ and $[Fe(CN)_6]^{4-}$ and 'Turnbull's blue' made from $Fe^{2+}$ and $[Fe(CN)_6]^{3-}$ pose the question as to whether they are essentially the same material. There is the possibility of different formulations such as $Fe_4^{3+}[Fe^{II}(CN)_6]_3$ and $Fe^{3+}Fe_3^{2+}[Fe^{III}(CN)_6]_3$, as well as the possibility of fast electron transfer to give an averaged electronic configuration. Their Mössbauer spectra are identical, and show quite elegantly that both are in fact $Fe_4^{3+}[Fe^{II}(CN)_6]_3$ [16]. Selective enrichment has also been used successfully in this case. Prussian blue prepared with enriched $^{57}Fe_2(SO_4)_3$ and Turnbull's blue prepared with enriched $^{57}FeCl_2$ both give a spectrum dominated by the $Fe^{3+}$ components. The formation of Turnbull's blue does not, therefore, involve any breaking of cyanide linkages, but is merely an electron transfer process. There is no substantial exchange between the cations and anions, which would have resulted in the $^{57}Fe$ apportioning between both.

Cyanide linkage isomerism has also been found to occur in the solid state in an iron-chromium complex [17]. The mixing of solutions of $K_3Cr(CN)_6$ and $FeSO_4$ generates a light brown precipitate (A) which approximates in composition to $Fe_{1.6}[Cr(CN)_6](OH)$ with an additional amount of water which cannot be removed by heating under vacuum. The Mössbauer spectrum shows all the iron to be present as $Fe^{2+}$ ions (the chemical isomer shift of ~1.25 mm $s^{-1}$ is characteristic) and the two partially resolved doublets with quadrupole splittings of Δ = 2.92 and 2.11 mm $s^{-1}$ have intensities in the ratio 0.6/1.0. This agrees with prediction from the formula assuming 37% of the $Fe^{2+}$ to be in interstitial sites and the remain-

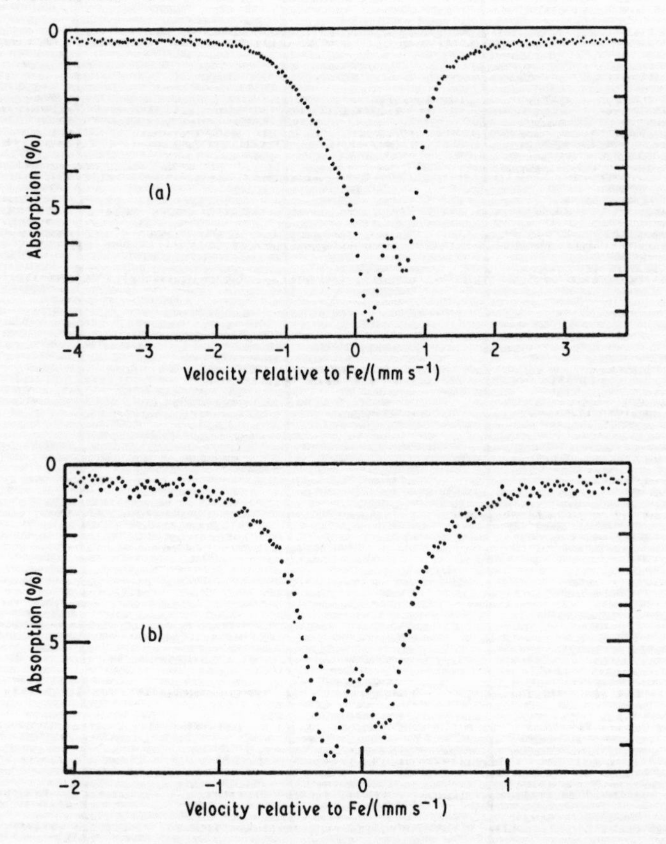


Fig. 3.4 $^{57}$Fe Mössbauer spectra at 77 K for Fe[Fe(CN)$_6$]. Spectrum (a) is for the normal material. Spectrum (b) is for a preparation using $^{57}$Fe enriched $K_3Fe(CN)_6$, and shows the quadrupole splitting (on an expanded scale) of the $[Fe^{III}(CN)_6]^{3-}$ anion. The $Fe^{3+}$ lines are now too weak to influence the spectrum. ([15], Figs. 9, 10).

der in the sites with co-ordination to nitrogen.

Mild heating generates a new darker brown complex (B). The lines from interstitial $Fe^{2+}$ disappear (although the $Fe^{2+}$ in nitrogen sites remains), and are replaced by a single absorption characteristic of a hexacyanoferrate(II). The interstitial $Fe^{2+}$ atoms have therefore interchanged with the $Cr^{3+}$ in the sites co-ordinated to carbon.

Heating above 80°C gives a dark green complex (C) which requires both oxygen and water for its formation. Its structure is still enigmatic, although the indications are of low-spin $Fe^{II}$ in carbon sites with oxidized high-spin $Fe^{3+}$ in nitrogen holes *or* interstitial sites.

Reduction of (C) with hydrazine hydrate causes a reversion to the original brown colour, but this product (D) is unstable in air. The Mössbauer spectrum shows that all of the iron not in the carbon sites has been reduced to $Fe^{2+}$, with a quadrupole splitting of ~2.8 mm $s^{-1}$ similar to the value already shown for interstitial sites. The overall reaction scheme can be most conveniently summarized as

| | (Carbon sites) | (Nitrogen sites) | (Interstitial sites) |
|---|---|---|---|
| A | $Cr^{III}$ | $Fe^{2+}$ | 0.6 $Fe^{2+}$ |
| B | 0.4 $Cr^{III}$, 0.6 $Fe^{II}$ | $Fe^{2+}$ | 0.6 $Cr^{3+}$ |
| C | [$Fe^{II}$ | (0.6 $Fe^{3+}$, $Cr^{3+}$) ?] | |
| D | $Fe^{II}$ | $Cr^{3+}$ | 0.6 $Fe^{2+}$ |

The reduced complex is apparently the true isomerization product with $Cr^{3+}$ in nitrogen holes, $Fe^{2+}$ in the carbon holes and the original amount of $Fe^{2+}$ in interstitial sites. However, it is convincingly shown that the linkage isomerism does not take place by a simple cyanide inversion, but by a complicated series of cation migrations.

## 3.4 Conformations in organometallic compounds of Fe

Many large molecules have considerable freedom of movement of the atoms within the structure, and several configurations may exist which are energetically very similar. However, the energy barrier between these may be very small so that interchange between the various 'conformations' can take place very easily. The cyclooctatetraene compound (COT)$Fe(CO)_3$ provides an interesting example of the way in which structural conformation can be established. The proton N.M.R. spectrum of the compound in various solvents at room temperature shows all eight protons to be equivalent; the obvious but incorrect inference is that the $C_8H_8$ ring is planar and bonding symmetrically to the $Fe(CO)_3$ moiety.

The X-ray crystal structure of the pure solid reveals asymmetric bonding of the iron to a $C_4H_4$ section of the ring, which is in fact

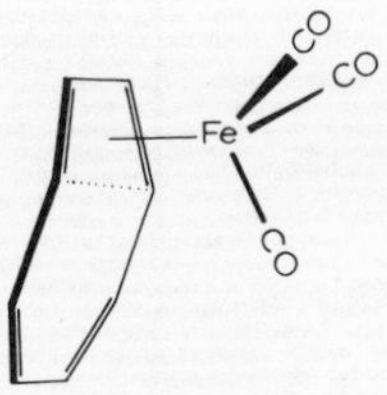

14

biplanar (structure 14) [18]. It can be shown that the room temperature N.M.R. spectrum results from a fast averaging effect by cooling the solutions to about 120 K, where the spectrum becomes more complicated as the molecular motion slows down. Clearly the molecule can adopt more than one conformation but with only a very small energy barrier between them which cannot prevent their interconversion at room temperature. However, three different interpretations of the nature of the low-temperature 'static' form of the complex have been given: a biplanar COT ring acting as a 1,3-diene as in the solid state [19]; a tub-shaped COT ring acting as a 1,5-diene [20]; and a fast isomerization in which the $Fe(CO)_3$ moiety moves between two equivalent structures in the 1,3-diene and 5,7-diene positions of a tub-shaped COT ring [21].

Although the Mössbauer effect cannot be observed in a solute molecule in the liquid phase, it is comparatively easy to freeze a solution into a glassy matrix where the individual molecules are not subject to the same nearest neighbour interactions as in the pure solid. It is imperative of course that the solute should not precipitate during freezing, and that the solvent should be both chemically inert and form a glass rather than a well-defined crystalline matrix.

Freezing $(COT)Fe(CO)_3$ in several solvents is seen to have a negligible effect on the Mössbauer parameters (Table 3.5, [22]). By way of contrast, the $Fe(CO)_3$ complex with cycloocta-1,5-diene and the 1,4-diene with norbornadiene show substantially different parameters (Table 3.5), and it is clear that the conjugated 1,3-diene structure found in the solid is retained in the low-temperature solution.

There are also examples where conformational changes *are* seen in solution. The compound $(\pi\text{-}C_5H_5)Fe(CO)_2SnCl_3$ shows the

Table 3.5 $^{57}Fe$ Mössbauer parameters in complexes with cyclooctatetraene

| Compound | Δ* | δ* |
|---|---|---|
| $(COT)Fe(CO)_3$ | 1.24 | +0.05 |
| in EPA solution† | 1.16 | +0.08 |
| in nitrobenzene | 1.21 | +0.10 |
| in cyclooctane | 1.23 | +0.10 |
| in methyltetrahydrofuran | 1.20 | +0.09 |
| cycloocta-1,5-diene $Fe(CO)_3$ | 1.83 | −0.03 |
| norbornadiene $Fe(CO)_3$ | 2.15 | +0.03 |

* in mm $s^{-1}$. δ is with respect to iron metal
† EPA is 16/42/42 % v/v ethanol/1-propanol/diethyl ether COT = cyclooctatetraene.

same $^{57}Fe$ and $^{119}Sn$ spectra in both the solid state and dispersed in a glassy polymethylmethacrylate matrix. The Mössbauer parameters are not influenced by the nature of the matrix. The $^{57}Fe$ spectra of $[(\pi\text{-}C_5H_5)Fe(CO)_2]_2SnCl_2$ are also uninfluenced by the matrix, but the $^{119}Sn$ spectrum shows four lines in the polymethylmethacrylate matrix as against only two in the solid state [23]. This is illustrated in Fig.3.5. Assignment of the 4-line spectrum into two quadrupole doublets can be made in more than one way, but the ambiguity does not affect the prime result, that the $^{119}Sn$ can have two different environments. Conformational effects have also been reported in solutions of $[(\pi\text{-}C_5H_5)Fe(CO)_2]SiCl_2CH_3$ and $[(\pi\text{-}C_5H_5)Fe(CO)_2]SnCl_2CH_3$, and it would appear that in the present instance two rotational conformers exist related by a 120° rotation about an Sn-Fe bond (structures 15–16). Structure 16 is that found in the solid state.

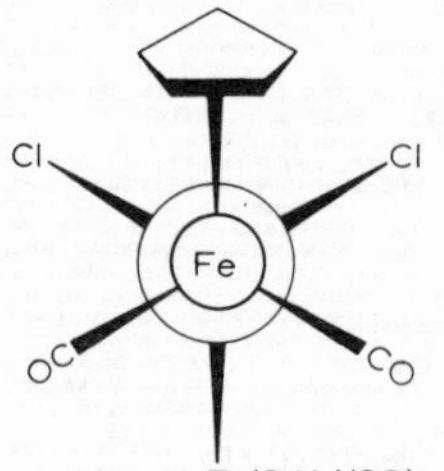

15

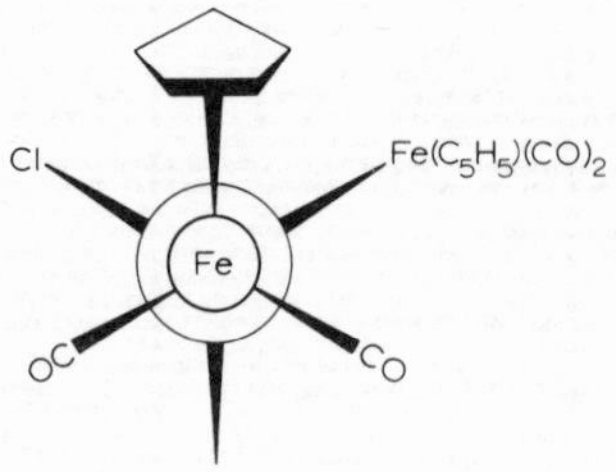

16

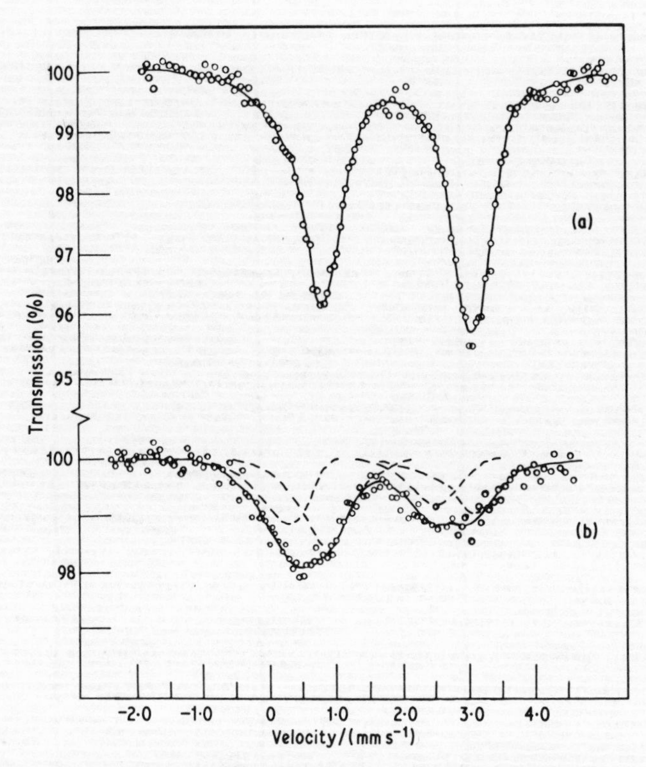


Fig. 3.5 The $^{119}Sn$ Mössbauer spectra of $[(\pi\text{-}C_5H_5)Fe(CO)_2]_2SnCl_2$ in the solid state (a) and in a polymethylmethacrylate matrix (b). The extra lines in the latter result from the presence of two rotational conformers in solution. ([23], Fig.2).

## 3.5 Stereochemistry in tin compounds

The study of organo-tin compounds has become an important area of research in inorganic chemistry. The $^{119}Sn$ Mössbauer spectrum has been recorded for a large number of organo-derivatives, but considerable difficulties have been experienced in interpreting the observed values of the chemical isomer shift and the quadrupole splitting. For example, it is disturbing to find no detectable quadrupole splitting in many compounds such as $Me_3SnPh$ and $Me_3SnH$ which are clearly unsymmetrical about the tin atom, whereas $Me_3SnCl$ and $Me_3SnOH$ for example show large splittings. There appears to be a clear distinction between bonds to carbon, and bonds to halides, oxygen, nitrogen or sulphur; only when the co-ordinated ligands are a mix of the two types is a large splitting observed. There is no obvious correlation with electronegativity, and current evidence favours a difference in $\sigma$-bonding characteristics which is not yet fully understood. In consequence, the lack of a quadrupole splitting is not necessarily indicative of a high symmetry in the tin environment.

A major complication has been the realization that many compounds which nominally have four co-ordinated ligands such as $Me_2SnCl_2$, $Me_3SnF$ and $Me_3SnCN$ do in fact feature a higher co-ordination (5 or 6) in the solid state as a result of one or more ligands being in a bridging position between two tin atoms. In the few cases where X-ray structures are known, unexpected deviations from regular geometry are sometimes found. Rationalization of the Mössbauer parameters in terms of stereochemistry is therefore not easy. A correspondence between geometry and the magnitude of the quadrupole splitting has been formulated on the basis of point-charge calculations (see p.66), but unfortunately does not allow an unambiguous assignment. It has also been proposed that polymeric bridged structures should have a significantly higher Debye temperature (and recoilless fraction) than purely monomeric structures, but again the differences are not sufficiently distinct for reliable diagnostic application.

The two parameters of the chemical isomer shift and quadrupole splitting are insufficient to make clear distinctions as to the nature of the geometrical environment in many instances. Extra information *can* be obtained by applying an external magnetic field with a flux density of the order of 3–5 $T$ to determine the sign of $e^2qQ$. Since $Q$ is negative for $^{119}Sn$, $e^2qQ$ is positive for $p_z$, $d_{z^2}$, $d_{xz}$ and $d_{yz}$ electrons, and negative for $p_x$, $p_y$, $d_{x^2-y^2}$ and $d_{xy}$. The applied field spectrum has a complex shape, but the sign can be determined by visual comparison with the spectrum predicted by theoretical calculation [24]. A typical example is shown in Fig.3.6. Although it would be useful to know the value of $\eta$, the method as applied to $^{119}Sn$ is not very sensitive to values of the asymmetry parameter of less than 0.5. However, a knowledge of the sign of $e^2qQ$ alone tells us something about the electrons producing $q$ and can frequently allow prediction of the stereochemistry.

$Me_2SnF_2$ and $Me_2SnCl_2$ which have a bridged-octahedral stereochemistry in the solid state with *trans*-axial methyl groups both give $e^2qQ$ as positive [25]. The sign of $e^2qQ$ for orbitals bonding

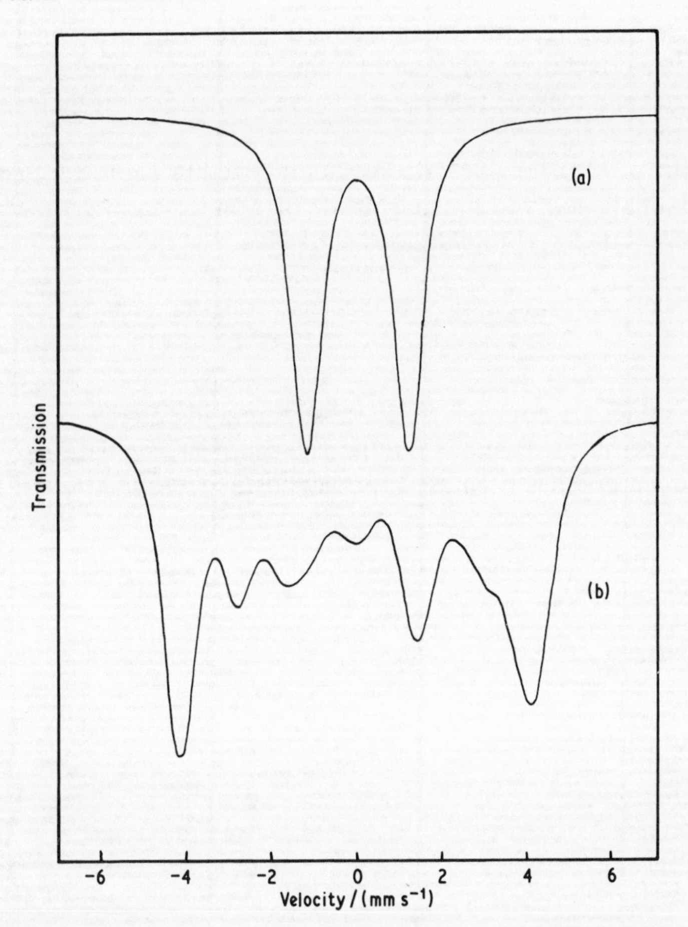


Fig. 3.6 Computer simulated spectra for a polycrystalline sample giving a quadrupole split $^{119}$Sn spectrum with $e^2qQ/2 = 2.215$ mm s$^{-1}$, $\eta = 0.6$ and a linewidth of 0.8 mm s$^{-1}$: (a) in zero field; (b) with a magnetic field with a flux density of 5 T (50 kG) applied perpendicular to the direction of observation. If $e^2qQ$ were to be negative in sign, the spectrum would be the mirror image of that shown.

in the *z* direction is also positive, consistent with the greater electron density being in the less-ionic Sn-C bonds, and since this situation may be anticipated to be general, we expect any similar *trans*-octahedral structures to also show a positive sign. $Me_3SnNCS$ and $Me_3SnOH$ which are known to have a trigonal bipyramidal geometry in the solid state with axial bridging NCS and OH groups respectively have a negative value of $e^2qQ$ (consistent with asymmetric σ-bonding, rather than π-donation from the bridging groups, or lattice effects). $Ph_3SnF$ and $Ph_3SnCl$ also have a negative sign for $e^2qQ$ and may be presumed to have similar 5-coordinate structures.

Some of the available data are summarized in Table 3.6. As an approximate guide to the magnitude of Δ in different geometries, simple point charge calculations [26] give

$$\begin{array}{ccccccccc} RSnX_5 & : & trans\text{-}R_2SnX_4 & : & cis\text{-}R_2SnX_4 & : & R_3SnX_2 & : & R_3SnX \\ +2 & : & +4 & : & -2 & : & -(3\text{ to }4) & : & -2 \end{array}$$

These values are approximately those found in practice, although differences in the R and X groups also cause large variations.

The method has been used for diagnostic determination of the

Table 3.6 $^{119}Sn$ Mössbauer parameters for organotin compounds

| Compound | Sign of $e^2qQ$ | Δ /(mm s$^{-1}$) | Stereo-chemistry | Ref. |
|---|---|---|---|---|
| $(Me_4N)_2[EtSnCl_5]$ | + | 1.94 | octahedral | 26 |
| $Pr^nSnCl_3$.2 piperidine | + | 1.99 | | 27 |
| $Me_2SnF_2$ | + | 4.65 | octahedral *trans*-alkyl | 25 |
| $Me_2SnCl_2$ | + | 3.4 | | 28 |
| $Cs_2[Me_2SnCl_4]$ | + | 4.28 | | 26 |
| $K_2[Me_2SnF_4]$ | + | 4.12 | | 26 |
| $Me_2SnCl_2(pyO)_2$ | + | 4.1 | | 29 |
| $Pr^n_2SnCl_2$.2β-picoline | + | 3.99 | | 27 |
| $Me_3SnNCS$ | − | 3.77 | trigonal bipyramidal | 25 |
| $Me_3SnOH$ | − | 2.91 | | 25 |
| $Ph_3SnF$ | − | 3.62 | | 25 |
| $Ph_3SnCl$ | − | 2.51 | | 25 |
| $(Me_4N)[Me_3SnCl_2]$ | − | 3.31 | | 26 |
| $(Me_4N)[Ph_3SnCl_2]$ | − | 3.02 | | 26 |
| $Et_3SnCN$ | − | 3.17 | | 26 |
| $Ph_3SnCl$.piperidine | − | 2.95 | | 27 |
| $Me_3SnC_6H_5$ | − | 1.39 | tetrahedral | 26 |
| $Ph_3SnC_6F_5$ | − | 0.97 | | 26 |
| $Me_2Sn(oxin)_2$ | + | 2.06 | octahedral *cis*-alkyl | 26 |
| $Ph_2Sn(oxin)_2$ | + | 1.67 | | 26 |
| $Ph_2Sn(NCS)_2$.1,10-phenanthroline | + | 2.36 | | 26 |
| $Pr^n_2SnCl_2$.2 morpholine | + | 2.41 | | 27 |

stereochemistry of new compounds [27]. $Ph_3SnCl$.piperidine for example has $e^2qQ$ negative in sign as predicted for a trigonal bipyramidal structure with equatorial phenyl groups, and contrary to prediction for equatorial Cl and piperidine ligands. $Pr^n_2SnCl_2$.2β-picoline evidently has *trans*-propyl groups. However, although $Pr^n_2SnCl_2$.2 morpholine has a quadrupole splitting approximately half that of the β-picoline compound, leading one to suspect a *cis*-geometry, the sign of $e^2qQ$ is positive in both cases and contrary to the arguments given earlier in the chapter regarding *cis-/trans*-isomerism. Moreover, other *cis*-complexes such as $Me_2Sn(oxin)_2$ for which an X-ray structure is available also show a positive value of $e^2qQ$. The problem here is that chelating ligands cause substantial distortion of the octahedral bond angles. The result is that the principal direction of the electric field gradient tensor has to be redefined and the value of $\eta$ may be quite large. In this example it is therefore partly fortuitous, although convenient, that the 2:1 *trans/cis* ratio described earlier still holds.

### 3.6 Molecular iodine compounds

Several good examples are available of the determination of geometry in an iodine compound. The covalently bonded molecule $I_2Cl_6$ has a planar bridged structure with two identical 4-coordinated iodine atoms (structure 17). The $^{129}I$ resonance gives a complex quadrupole pattern from the $\frac{5}{2} \rightarrow \frac{7}{2}$ decay. An example is shown on p.81. The derivative compound $I_2Cl_4Br_2$ can have several hypothetical structures (18a-18f). The $^{129}I$ spectrum of this compound [30] shows two superimposed quadrupole patterns of equal intensity, indicative of two dissimilar iodine atoms. This eliminates symmetrical structures b, c and d. Closer inspection reveals that one iodine atom is apparently identical to iodine in $I_2Cl_6$, which eliminates structures e and f and establishes structure a, in which both bromine atoms are terminally substituting on the same iodine atom, to be the correct one.

The quadrupole splitting of $^{129}I$ in complexes of ICl and IBr with pyridine bases, e.g. ICl.pyridine, has a negligible asymmetry parameter [31]. This is good evidence for collinearity of the Cl-I-N bonds in these compounds.

The $^{129}I$ spectrum of $[IF_6]^+[AsF_6]^-$ is a single sharp line (Fig. 3.7). This confirms the regular octahedral symmetry of the $IF_6^+$

17

(a) (b)

(c) (d)

(e) (f)

18

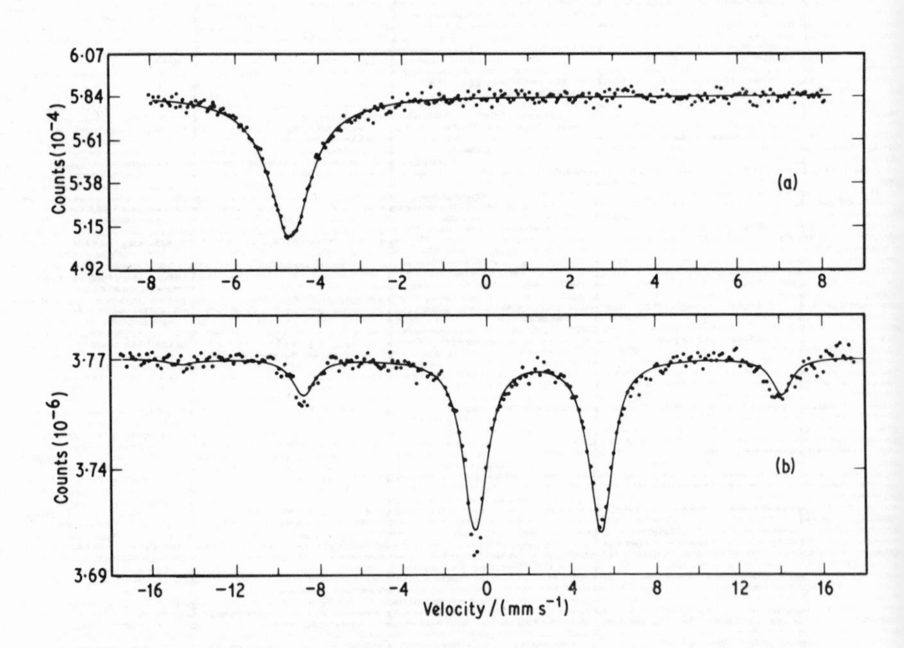


Fig. 3.7 The $^{129}$I spectra of (a) the $IF_6^+$ cation and (b) the $IF_6^-$ anion at 90 K. The single line in the former case is consistent with regular octahedral symmetry, while the quadrupole splitting with a large asymmetry parameter for $IF_6^-$ indicates a stereochemically active lone-pair of electrons. ([32], Fig. 1).

19

cation [32], and contrasts strongly with the large quadrupole splitting found for the $IF_6^-$ anion in $CsIF_6$. The spectrum in this case is nearly symmetrical due to a large asymmetry parameter, $\eta$, and indicates that the additional lone-pair of electrons is stereochemically active. A suggested structure is derived from the pentagonal bipyramid of $IF_7$ by replacing an equatorial fluorine with the lone-pair (structure 19).

## Appendix

### *Quadrupole splitting in* cis- *and* trans-*isomers*

The electric field gradient at the nucleus is derived from the second derivative of the electric potential due to all the extranuclear charges, and in particular to valence electrons and neighbouring ions. At a naïve level, a complex $MAB_5$ is considered as comprising a charge $q_A$ at distance $R$, and five charges $q_B$ at distance $r$ from M.

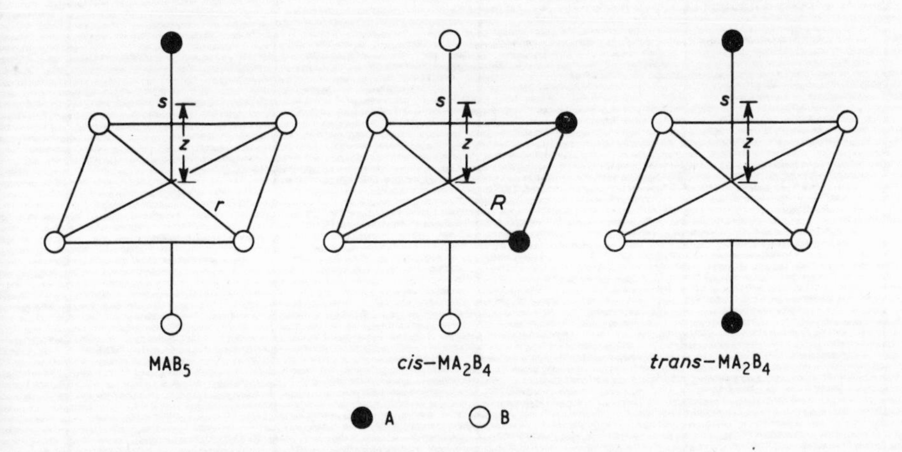

The *cis*- and *trans*-isomers of $MA_2B_4$ can be treated similarly, and the principal axis of the electric field gradient tensor is indicated in each case in the figure as the $z$ axis.

Considering $MAB_5$, the potential at point s due to $q_A$ is given by

$$V_s^A = \frac{q_A}{R - z}$$

thence

$$\frac{dV_s^A}{dz} = \frac{q_A}{(R - z)^2}$$

and

$$\frac{d^2V_s^A}{dz^2} = \frac{2q_A}{(R - z)^3}$$

The value as $z \to 0$ becomes

$$\frac{d^2V^A}{dz^2} = \frac{2q_A}{R^3}$$

which is the contribution of charge $q_A$ to $V_{zz}$. Including all terms in the potential,

$$V_s = \frac{q_A}{(R - z)} + \frac{q_B}{(r + z)} + \frac{4q_B}{(r^2 + z^2)^{1/2}}$$

$$\frac{d^2V_s}{dz^2} = \frac{2q_A}{(R - z)^3} + \frac{2q_B}{(r + z)^3} - \frac{4q_B(r^2 - 2z^2)}{(r^2 + z^2)^{5/2}}$$

and as $z \to 0$

$$V_{zz}(mono) = \frac{2q_A}{R^3} - \frac{2q_B}{r^3}$$

Similarly for *cis*-$MA_2B_4$

$$V_s = \frac{2q_A}{(R^2 + z^2)^{1/2}} + \frac{2q_B}{(r^2 + z^2)^{1/2}} + \frac{q_B}{(r + z)} + \frac{q_B}{(r - z)}$$

hence

$$V_{zz}(cis) = -\frac{2q_A}{R^3} + \frac{2q_B}{r^3} = -V_{zz}(mono)$$

Likewise for *trans*-$MA_2B_4$

$$V_s = \frac{q_A}{(R + z)} + \frac{q_A}{(R - z)} + \frac{4q_B}{(r^2 + z^2)^{1/2}}$$

$$V_{zz}(trans) = \frac{4q_A}{R^3} - \frac{4q_B}{r^3} = -2V_{zz}(cis)$$

Thus we have the result

$$2V_{zz}(mono) = -2V_{zz}(cis) = V_{zz}(trans)$$

The result is valid as long as the bonds are basically similar in all three compounds, and is independent of the magnitudes of $q_A$ and $q_B$. However, if for example there were a stronger interaction between the A groups when *trans*- to each other, then deviations may be expected.

## References

[1] Gibb, T. C. (1970) In *Spectroscopic Methods in Organometallic Chemistry*, Chapter 2, p.47, ed. W.O. George, Butterworths, London.
[2] Farmery, K., Kilner, M., Greatrex, R. and Greenwood, N. N. (1968) *Chem. Comm.*, **593**.
[3] Gibb, T. C., Greatrex, R., Greenwood, N. N. and Thompson, D. T. (1967) *J. Chem. Soc. (A)*, 1663.
[4] Clark, M. G., Cullen, W. R., Garrod, R. E. B., Maddock, A. G. and Sams, J. R. (1973) *Inorg. Chem.*, **12**, 1045.
[5] Herber, R. H., Kingston, W. R. and Wertheim, G. K. (1963) *Inorg. Chem.*, **2**, 153.
[6] Greatrex, R., Greenwood, N. N., Rhee, I., Ryang, M. and Tsutsumi, S. (1970) *Chem. Comm.*, 1193.
[7] Dahl, L. F. and Rundle, R. E. (1957) *J. Chem. Phys.*, **26**, 1751.

[8] Erickson, N. E. and Fairhall, A. W. (1965) *Inorg. Chem.*, **4**, 1320.
[9] Wei, C. H. and Dahl, L. F. (1969) *J. Amer. Chem. Soc.*, **91**, 1351.
[10] McDonald, W. S., Moss, J. R., Raper, G., Shaw, B. L., Greatrex, R. and Greenwood, N. N. (1969) *Chem. Comm.*, 1295.
[11] Cullen, W. R., Harbourne, D. A., Liengme, B. V. and Sams, J. R. (1969) *Inorg. Chem.*, **8**, 95.
[12] Berrett, R. R. and Fitzsimmons, B. W. (1966) *Chem. Comm.*, **91**; (1967) *J. Chem. Soc. (A)*, 525.
[13] Bancroft, G. M., Garrod, R. E. B., Maddock, A. G., Mays, M. J. and Prater, B. E. (1970) *Chem. Comm.*, 200.
[14] Fitzsimmons, B. W., Seeley, N. J. and Smith, A. W. (1969) *J. Chem. Soc. (A)*, 143.
[15] Maer, K., Beasley, M. L., Collins, R. L. and Milligan, W. O. (1968) *J. Amer. Chem. Soc.*, **90**, 3201.
[16] Kerler, W., Neuwirth, W., Fluck, E., Kuhn, P. and Zimmermann, B. (1963) *Z. Physik*, **173**, 321.
[17] Brown, D. B., Shriver, D. F. and Schwartz, L. H. (1968) *Inorg. Chem.*, **7**, 77.
[18] Dickens, B. and Lipscomb, W. N. (1962) *J. Chem. Phys.*, **37**, 2084.
[19] Kreiter, C. G., Maasbol, A., Anet, F. A. L., Kaesz, H. D. and Winstein, S. (1966) *J. Amer. Chem. Soc.*, **88**, 3444; Anet, F. L., Kaesz, H. D., Maasbol, A. and Winstein, S. (1967) *J. Amer. Chem. Soc.*, **89**, 2489.
[20] Cotton, F. A., Davidson, A. and Faller, J. W. (1966) *J. Amer. Chem. Soc.*, **88**, 4507.
[21] Keller, C. E., Shoulders, B. A. and Pettit, R. (1966) *J. Amer. Chem. Soc.*, **88**, 4706.
[22] Grubbs, R., Breslow, R., Herber, R. H. and Lippard, S. J. (1967) *J. Amer. Chem. Soc.*, **89**, 6864.
[23] Herber, R. H. and Goscinney, Y. (1968) *Inorg. Chem.*, **7**, 1293.
[24] Gibb, T. C. (1970) *J. Chem. Soc. (A)*, 2503.
[25] Goodman, B. A. and Greenwood, N. N. (1971) *J. Chem. Soc. (A)*, 1862.
[26] Parish, R. V. and Johnson, C. E. (1971) *J. Chem. Soc. (A)*, 1906.
[27] Goodman, B. A., Greenwood, N. N., Jaura, K. L. and Sharma, K. K. (1971) *J. Chem. Soc. (A)*, 1865.
[28] Goodman, B. A. and Greenwood, N. N. (1969) *Chem. Comm.*, 1105.
[29] Fitzsimmons, B. W. (1970) *J. Chem. Soc. (A)*, 3235.
[30] Pasternak, M. and Sonnino, T. (1968) *J. Chem. Phys.*, **48**, 1997.
[31] Wynter, C. I., Hill, J., Bledsoe, W., Shenoy, G. K. and Ruby, S. L. (1969) *J. Chem. Phys.*, **50**, 3872.
[32] Bukshpan, S., Soriano, J. and Shamir, J. (1969) *Chem. Phys. Letters*, **4**, 241.

CHAPTER FOUR

# *Electronic Structure and Bonding: Diamagnetic Compounds*

Once the molecular structure of a compound has been established, it becomes possible to study the nature of the chemical bonding in more detail. Mössbauer data now assume a more important role, because once the number and disposition of the groups bonding to a resonant nucleus are known, the hyperfine interactions can be used to investigate the electron distribution in individual bonds.

The important hyperfine effects were introduced in Chapter 2. The chemical isomer shift is a direct function of the *s*-electron density at the nucleus, but shows secondary induced effects due to changes in shielding of the *s*-electrons by *p*-, *d*- or *f*-electrons. It is therefore sensitive to most changes of orbital occupation in the valence shell of the atom. The quadrupole splitting is derived from asymmetric occupation of the non-spherical *p*-, *d*- or *f*-electron orbitals, and its magnitude and sign can be used to determine the directional orientation of the asymmetry with respect to the chemical bonds. Diamagnetic compounds do not show a magnetic hyperfine splitting unless a large external magnetic field is applied. The generation of a large *internal* magnetic field at the nucleus by an unpaired electron will be described in the next chapter on paramagnetic compounds.

The presence of the unpaired electrons in paramagnetic compounds results in a wide variety of effects, many of which are strongly temperature dependent. In contrast, the electronic properties of diamagnetic compounds are largely unaffected by temperature variation, resulting in a significant reduction in the amount of information obtainable from the hyperfine interactions. The primary observable parameters become the chemical isomer shift and the quadrupole splitting, and these are both intimately linked to the covalent bonding of the molecule.

Because of this distinction, it is convenient to discuss covalency in diamagnetic substances in some depth before turning to some of the more exotic phenomena associated with unpaired electrons. As an introduction, it is instructive to consider, simply, the effects of a change in the formal oxidation state of the resonant atom.

## 4.1 Formal oxidation state

Some intimation of the effect on the Mössbauer spectrum of a change in the formal oxidation state of the resonant atom was given by several of the examples in Chapter 3. A change in oxidation state involves the addition or removal of a valence electron, and manifests itself as a change in the chemical isomer shift. This is true of both diamagnetic and paramagnetic electron configurations. For example, the oxidation of tin(II) to tin(IV) formally involves the loss of two 5*s*-electrons which will result in a large change in the $^{119}Sn$ chemical isomer shift. Similarly, the oxidation of $Fe^{2+}$ to $Fe^{3+}$ involves loss of a 3*d*-electron and causes a reduction in the shielding of the outer *s*-electrons.

Accurate estimation of the changes in electron-density is difficult. The problems involved can be illustrated by considering the different oxidation states of iron. Figures are available for several free-ion configurations of iron, based on restricted Hartree-Fock calculations by Watson [1, 2], and show the effect of a decreasing number of 3*d*-electrons on the value of $|\psi_s(0)|^2$. As may be seen from Table 4.1, the largest effect is felt in the contribution from $|\psi_{3s}(0)|^2$. A progressive reduction in the number of 3*d*-electrons causes a non-linear increase in the total *s*-electron density.

Unfortunately, the value of the fractional change in the nuclear radius in equation (2.4), $\delta R/R$, cannot be determined independently of the chemical environment. It is therefore necessary to construct an empirical calibration of the chemical isomer shift scale using the observed shifts of compounds whose electron configurations are known. In this way the sign of $\delta R/R$ and its approximate magnitude are obtained. However, no known compound of iron can be assigned the free-ion configuration $3d^5$ or $3d^6$ because all $Fe^{3+}$ and

Table 4.1 Electron densities at $r = 0$ for different configurations of an Fe atom [1, 2]

| Electrons per cubic Bohr radius | $3d^8$ | $3d^7$ $Fe^+$ | $3d^6$ $Fe^{2+}$ | $3d^5$ $Fe^{3+}$ | $3d^64s^2$ free atom |
|---|---|---|---|---|---|
| $\lvert\psi_{1s}(0)\rvert^2$ | 5378.005 | 5377.973 | 5377.840 | 5377.625 | 5377.873 |
| $\lvert\psi_{2s}(0)\rvert^2$ | 493.953 | 493.873 | 493.796 | 493.793 | 493.968 |
| $\lvert\psi_{3s}(0)\rvert^2$ | 67.524 | 67.764 | 68.274 | 69.433 | 68.028 |
| $\lvert\psi_{4s}(0)\rvert^2$ | | | | | 3.042 |
| $2\sum_n \lvert\psi_{ns}(0)\rvert^2$ | 11878.9 | 11879.2 | 11879.8 | 11881.7 | 11885.8 |

$Fe^{2+}$ compounds show some measure of covalent character. The classic work on the calibration of the $^{57}Fe$ chemical isomer shift by Walker, Wertheim and Jaccarino [3], although valid in principle, is incorrect in detail because it neglects the considerable 4*s*-character in the covalent bonding of the $Fe^{3+}$ compounds.

A more realistic approach is to carry out a full molecular-orbital calculation on a model compound such as the $FeF_6^{3-}$ cluster, and to equate the derived electron-densities with the observed shift [4]. A newer and alternative method is to isolate atoms and ions of the element in an inert matrix such as solid xenon. Experiments which have produced neutral iron atoms in the $3d^64s^2$ configuration and $Fe^+$ ions in the $3d^64s^1$ and $3d^7$ configurations are described in Chapter 10. However, in all instances the results are dependent on the accuracy of the theoretical wave-functions used.

Despite these difficulties it is comparatively easy to determine the relative change in *s*-electron density between oxidation states and thence the sign of $\delta R/R$. Thus the $Fe^{2+}$ oxidation state gives a chemical isomer shift at higher velocity than $Fe^{3+}$; Table 4.1 shows that the *s*-electron density at the nucleus is smaller in $Fe^{2+}$, so that $\delta R/R$ for the 14.4-keV transition in $^{57}Fe$ is negative in sign.

Although each oxidation state has a different chemical isomer shift, within any given oxidation state the shift can vary from compound to compound because of the effects of covalent character already alluded to. The approximate ranges for iron, compiled empirically from experimental data, are shown in Fig.4.1. This figure also illustrates an important secondary consideration; name-

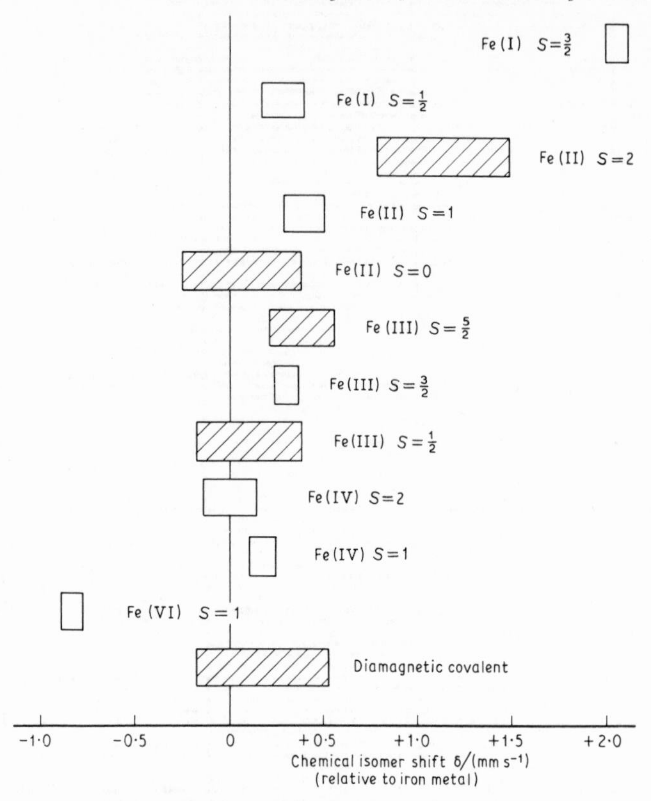


Fig. 4.1 Approximate representation of the ranges of chemical isomer shifts found in iron complexes. The values are related to iron metal at room temperature, but do allow for temperature variations in the absorber. The width of the individual distributions may be attributed to covalency, and as a general rule the more ionic compounds have the higher shift. The most common configurations are cross-hatched, and the diamagnetic covalent classification includes carbonyls and ferrocenes etc. In most series the shift decreases as spin-pairing takes place. Some of the less common configurations may in fact extend beyond the limits shown. Note that the Fe(IV) $S = 2$ subgroup refers to oxides with collective-electron bonding which may account for their anomalous position as regards the Fe(IV) $S = 1$ compounds which have a true low-spin configuration.

ly the effect of differing electron correlation. For example the $3d^6$ configuration of iron(II) can exist in high-spin ($S = 2$), intermediate-spin ($S = 1$), and low-spin ($S = 0$) forms. These differ substantially in their shielding effect on the *s*-electrons, and therefore show distinct chemical isomer shifts. An increase in covalent character may be expected to cause an effective decrease in 3*d*-shielding due to the nephelauxetic ('cloud-expanding') effect; thus the *s*-electron density increases and the chemical isomer shift decreases. For the more ionic compounds, the degree of covalent character can usually be estimated approximately from the chemistry of the atoms or groups bonding to the resonant nucleus, so that the chemical isomer shift can be predicted more closely than Fig.4.1 would seem to indicate.

The diagnostic use of the determination of the oxidation state has already been indicated in Fig.3.1, and in the discussion of linkage isomerism. Further examples will become apparent elsewhere. It is now also clear that differences in electronic configuration and covalent bonding produce characteristic effects. Similar arguments hold for resonant nuclei other than $^{57}Fe$, although the distinctions become less clear for elements with strong covalent bonding such as iodine. For example, tin(II) compounds give a $^{119}Sn$ resonant absorption between +2.9 and +4.5 mm $s^{-1}$ (relative to $BaSnO_3$) while the range for tin(IV) is –0.5 to +2.0 mm $s^{-1}$. Europium(II) compounds give a 21.6-keV $^{151}Eu$ absorption in the region –13.9 to –10.9 mm $s^{-1}$ (relative to $EuF_3$), compared to –0.2 to +0.9 mm $s^{-1}$ for europium(III). The large difference between $Eu^{2+}$ and $Eu^{3+}$ is due entirely to the reduced shielding of the *s*-electron core upon removal of an *f*-electron, and indicates $\delta R/R$ to be positive in sign.

The foregoing discussion has been based solely on deductions from the chemical isomer shift. In some instances where unpaired electrons are present, the quadrupole splitting can also be used to indicate the oxidation state. Examples of this are given in Chapter 5.

## 4.2 Iodine

The fortunate occurrence of suitable Mössbauer isotopes for the contiguous series tin, antimony, tellurium, iodine and xenon has greatly facilitated the study of chemical bonding in these elements. Their compounds are diamagnetic, and the orbitals participating most strongly in the bonding are the 5*s*- and 5*p*-orbitals, with some contribution from 5*d*- in appropriate instances. The ability of iodine to form compounds involving only one covalent bond to another atom makes it possible to develop a quantitative interpretation of the Mössbauer parameters which is not always possible for other elements. In this discussion of covalency, it is instructive to consider iodine first, before tackling the more difficult problems which occur elsewhere.

Data can be obtained from both the 27.7-keV $^{129}I$ and 57.6-keV $^{127}I$ resonances, the former giving better spectra (see Fig.2.1) but with the inconvenience of being an artificial radioactive isotope. Both levels are populated by the $\beta$-decay of a tellurium parent, so that chemical isomer shift data are usually quoted with respect to the customary source matrix, zinc telluride. The $^{129}I$ excited state spin of $I_e = \frac{5}{2}$ and ground state spin of $I_g = \frac{7}{2}$ are interchanged in $^{127}I$, but otherwise the analysis of the respective quadrupole spectra are equivalent. The energy-level scheme appropriate to $^{129}I$ is illustrated in Fig.2.3.

The chemical isomer shift for $^{127}I$ or $^{129}I$ can be expressed in terms of the deviation from the closed-shell electronic configuration $5s^2 5p^6$ of the $I^-$ anion [5, 6]. The nephelauxetic effect of covalent bonding causes an electron to become less effective in the hyperfine interactions. In particular, a $5p$-electron in a covalent bond has a very much smaller value of $\langle r^{-3} \rangle$ so that it appears to be a 'hole' in the closed-shell configuration. This concept has already been well proven in the interpretation of nuclear quadrupole resonance (N.Q.R.) data. If we define the number of $5s$- and $5p$-electron 'holes' in the $I^-$ configuration to be $h_s$ and $h_p$ respectively, the chemical isomer shift with respect to $I^-$ can be written as

$$\delta_A = K[-h_s + \gamma(h_p + h_s)(2 - h_s)]$$

where $K$ and $\gamma$ are constants. The parameter $K$ in fact contains $\delta R/R$ and is therefore numerically different for $^{127}I$ and $^{129}I$. The term $-Kh_s$ represents the decrease in $s$-electron density at the nucleus due to the loss of $5s$-electrons, the proportionality being assumed linear. The $(2 - h_s)$ remaining $5s$-electrons are then deshielded by a total decrease in the number of electrons of $h_p + h_s$. This assumes that the $5s$- and $5p$-electrons produce an equivalent shielding effect. In the event that $h_s$ is negligible, we have

$$\delta_A = 2K\gamma h_p$$

The value of $\gamma$ has been determined by calculation to be 0.07 [7], but the derivation of a numerical value for $K$ requires a more circumspect approach.

The relative occupation of the $5s$- and $5p$-orbitals is not given directly by the chemical isomer shift, but has to be derived in conjunction with the quadrupole splitting to which $h_s$ does not contribute. The relevant theory is a direct adaptation from nuclear quadrupole resonance spectroscopy [8]. The Townes and Dailey theory relates the principal value of the molecular field gradient ($eq_{\text{mol}}$) to the atomic electric field gradient from a $5p_z$-electron ($eq_{\text{at}}$) by

$$eq_{\text{mol}} = -eq_{\text{at}}U_p$$

where $U_p$ is the effective $5p$-electron imbalance. $U_p$ is defined in terms of the $5p$-electron populations in the $x$, $y$, and $z$ directions, $U_x$, $U_y$, and $U_z$ by

$$U_p = -U_z + \tfrac{1}{2}(U_x + U_y)$$

This follows directly because if the major axis of the electric field gradient tensor is defined as the $z$ axis, then $eq(p_z) = -\frac{1}{2}\,eq(p_x, p_y)$. The asymmetry parameter is given by

$$\eta = (V_{xx} - V_{yy})/V_{zz}$$

so that

$$\eta = \tfrac{3}{2}(U_x - U_y)/U_p$$

For axial symmetry, $U_x = U_y$ and $\eta = 0$. The value of $(e^2q_{\text{at}}{}^{127}Q_g/h)$ is known accurately from N.Q.R. work to be +2293 MHz, and it is customary to quote both $^{127}$I and $^{129}$I quadrupole coupling constants in the same units [the appropriate conversions from a quadrupole splitting in units of velocity, $v$, are obtained from $(v/c)(E_\gamma/h)$ and are: $^{127}$I 1 mm s$^{-1} \equiv$ 46.46 MHz, $^{129}$I 1 mm s$^{-1} \equiv$ 32.58 MHz]. $U_p$ can now be obtained directly by proportionality, and having calculated $U_z$, $U_x$ and $U_y$ from $e^2q^{127}Q_g$ and $\eta$, $h_p$ can be derived as

$$h_p = 6 - (U_x + U_y + U_z)$$

Calibration of the chemical isomer shift is made by using data for the iodine molecule isolated in frozen solutions (see below) in which $h_p = 0.99$, and for the iodide anion in which $h_p = 0$. This gives the equations

$$\delta(^{129}\mathrm{I}) \simeq -8.2h_s + 1.5h_p - 0.54 \text{ mm s}^{-1}$$

$$\delta(^{127}\mathrm{I}) \simeq +3.44h_s - 0.56h_p + 0.16 \text{ mm s}^{-1}$$

where in both cases the reference zero is the source matrix ZnTe. Note the inversion of sign caused by the opposing signs of the $\delta R/R$ values.

It can be seen from these equations that, in addition to permitting a determination of 5*p*-imbalance in the equivalent manner to N.Q.R. spectroscopy, the Mössbauer spectrum also allows a determination of the 5*s*-orbital occupation. The two figures in combination give a much more detailed description of the covalent bonding.

The $^{129}$I spectrum of solid iodine at 100 K is shown in Fig.4.2 and illustrates the good resolution which can be obtained with this isotope [9]. The large number of lines in the quadrupole split spectrum allows an unambiguous determination of both the value and sign of $e^2qQ$, and also of the asymmetry parameter $\eta$. The iodine atom in solid iodine has $e^2q^{127}Q_g/h = -2085$ MHz, but surprisingly has a non-zero asymmetry parameter of $\eta = 0.16$. This is unexpected for a linear molecule and is contrary to the naïve assumption that the bonding is purely of *p*-character. The results are consistent with admixture of, for example, an $s^2p^4d$ configuration into the $s^2p^5$ state as a result of intermolecular interactions. However, frozen solutions of iodine in hexane, $CCl_4$, and solid argon show a similar spectrum with $e^2q^{127}Q_g/h \simeq -2250$ MHz but here the asymmetry parameter is zero [10]. Clearly in the latter case a closer approach to a true isolated $I_2$ molecule is found, and the calculated $h_p$ value of ~0.99 is in good agreement with the predicted value for a free molecule of 1.00.

Having established the calibration of the isomer shift scale using $I^-$ and the $I_2$ molecule, it becomes feasible to study the chemical bonding in other compounds. Sample experimental data are collected in Table 4.2. Some of the numerical values for $U_p$ and $h_p$ differ from their original source because all data have been recalculated using the latest calibration figures.

The value of $e^2q^{127}Q_g/h$ in IBr is $-2892$ MHz, and since this

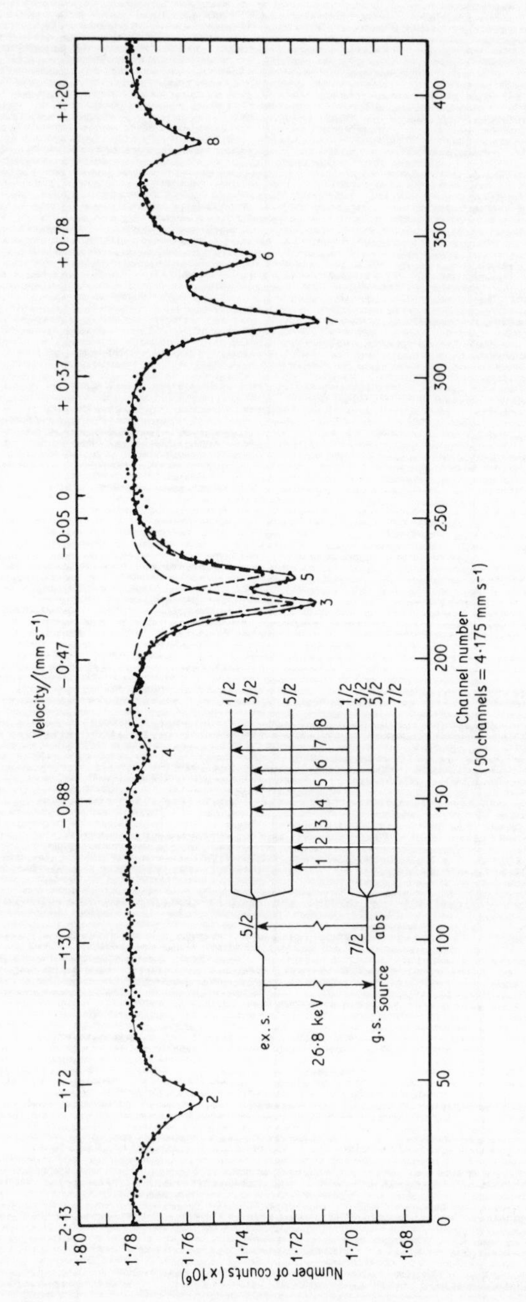


Fig. 4.2 The $^{129}$I resonance in solid $^{129}I_2$ at 100 K. The line numbered 1 on the decay scheme is off the scale of the spectrum. It is usually ignored because of its low intensity. ([9], Fig. 3).

Table 4.2 Mössbauer data for the $^{129}$I resonance

| Compound | $T$/K | $^{129}\delta$(ZnTe) /(mm s$^{-1}$) | $e^2q^{127}Q_g h^{-1}$ /MHz | $\eta$ | $U_p$ | $h_p$ | $h_s$ | Reference |
|---|---|---|---|---|---|---|---|---|
| $I_2$ (solid) | 100 | +0.82 | −2085 | 0.16 | +0.91 | – | – | 9 |
| $I_2$ (inert matrix) | 88 | +0.94 | −2250 | 0 | +0.99 | 1.00 | – | 10 |
| KI | 80 | −0.51 | 0 | 0 | – | 0.02 | – | 5 |
| ICl | 80 | +1.73 | −3131 | 0.06 | +1.38 | 1.48 | – | 11 |
| IBr | 80 | +1.23 | −2892 | 0.06 | +1.26 | 1.18 | – | 11 |
| ICN | 80 | +1.19 | −2640 | 0.00 | +1.15 | 1.15 | – | 12 |
| $CH_3I$ | 85 | +0.20 | −1739 | 0 | +0.76 | 0.76 | 0.04 | 13 |
| $CH_2I_2$* | 4.2 | +0.42 | −1920 | | | | | |
| $CHI_3$ | 85 | +0.53 | −2029 | 0 | +0.88 | 0.88 | 0.03 | 13 |
| $CI_4$ | 85 | +0.65 | −2102 | 0 | +0.91 | 0.91 | 0.02 | 13 |
| $SiI_4$ | 85 | +0.26 | −1335 | 0 | +0.58 | 0.53 | – | 15 |
| $GeI_4$ | 85 | +0.48 | −1500 | 0 | +0.65 | 0.68 | – | 15 |
| $SnI_4$ | 85 | +0.43 | −1364 | 0.00 | +0.59 | 0.64 | – | 14 |

* values estimated from $^{127}$I data.

is a linear molecule gives directly a value for $U_p$ of 1.26 [11]. If the bonding involves only $p_z$ electrons, then $h_p = 1.26$. Alternatively, using the chemical isomer shift calibration with $h_s = 0$ gives $h_p = 1.18$. The discrepancy is probably within the accuracy of the calibration, but it is noteworthy that it corresponds to a value for $h_s$ of only 0.01. We can therefore be confident that the *s*-character in the bonding is very small, and that ~0.25 $e^-$ is transferred from the iodine to bromine. Similarly in ICN 0.15 $e^-$ is transferred to the cyanide [12], and in ICl 0.48 $e^-$ is transferred to the chlorine [10]. In both IBr and ICl there is a small asymmetry parameter, and the observed discrepancies may well be the result of intermolecular interactions in the solid.

The iodomethanes $CH_3I$, $CH_2I_2$, $CHI_3$ and $CI_4$ are not expected to involve $\pi$-bonding, and $U_p$ then gives $h_p$ directly [13]. The *p*-electron transfer to iodine decreases in the order 0.24 $e^-$, 0.16 $e^-$, 0.12 $e^-$ and 0.09 $e^-$ per iodine as the number of substituent iodines increases, and at the same time the *s*-character in the bonding decreases. The sensitivity of the data to the *s*-character of the bond is shown in Fig.4.3 where the chemical isomer shift and quadrupole splitting values are plotted around the line calculated for pure 5*p*-bonding. The deviations shown by the iodomethanes are

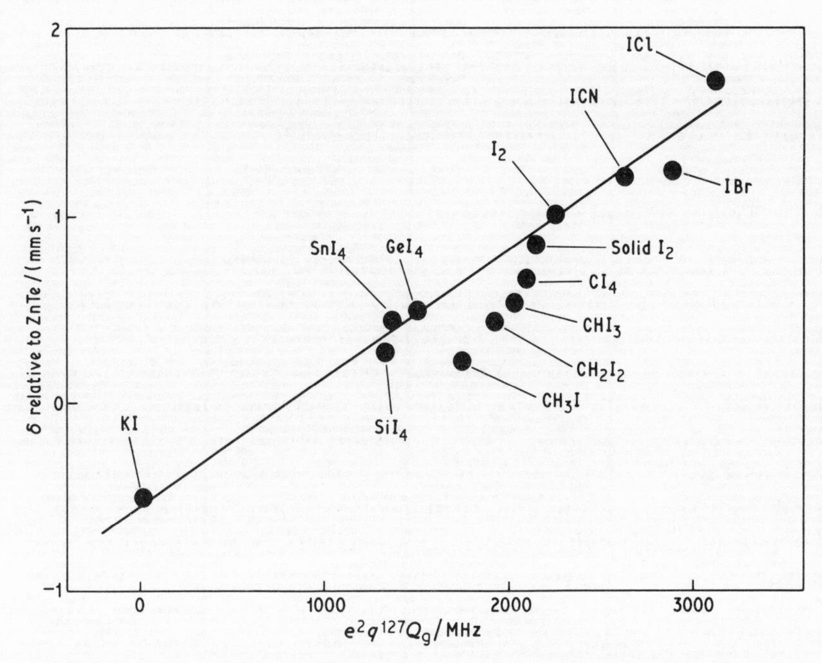


Fig. 4.3 Values for the chemical isomer shift and quadrupole coupling constant in iodine compounds plotted around the line calculated for pure 5*p*-bonding. Note the substantial deviations in the iodomethanes due to an s-character of less than 5%.

substantial, despite the fact that $CH_3I$ has only about 4% *s*-character.

Compounds in the series $CI_4$, $SiI_4$, $GeI_4$ and $SnI_4$ [14, 15] all feature essentially pure $p_\sigma$ bonding and show no direct evidence of appreciable *s*-character or $p_\pi$-character in the bonding; the amount of electron density transferred to iodine is 0.09, 0.47, 0.32 and 0.36 $e^-$ respectively.

Compounds which show chemical isomer shifts outside the range spanned by $I^-$ and $I_2$ are usually (if perhaps unwisely) interpreted by means of the same calibration. However, the large differences between for example $IF_5$ (+3.00 mm $s^{-1}$), $IF_7$ (−4.56), $IF_6^+$ (−4.68) and $IF_6^-$ (+2.45) make the estimated orbital occupations very sensitive to the numerical values of the constants in the calibration. It is also noteworthy that there is no obvious correspondence between the chemical isomer shift and the formal oxidation state. The large amount of *s*-character in the bonding of $IF_7$ and $IF_6^+$

causes apparently anomalous shifts. The difficulties arise mainly because of the large number of orbitals directly involved in bonding when the co-ordination is high. This problem becomes more acute in the elements immediately preceeding iodine.

## 4.3 Tellurium and antimony

The 35.5-keV resonance in $^{125}Te$ has an observed range of chemical isomer shifts which is less than the natural linewidth of 5.02 mm $s^{-1}$, and the quadrupole splittings from the $I_g = \frac{1}{2} \rightarrow I_e = \frac{3}{2}$ transition are also poorly resolved. Comparatively few data for inorganic compounds are available, but several important generalizations have emerged [16]. The octahedral tellurium(IV) complexes such as $TeCl_6^{2-}$, $TeBr_6^{2-}$ and $TeI_6^{2-}$ show the most positive chemical isomer shifts. They are unique in having a stereochemically inactive pair of non-bonding electrons which may be presumed to have considerable 5*s*-character, and to be highly contracted towards the Te nucleus. Other tellurium(IV) compounds such as square pyramidal $TeF_5^-$, pyramidal $TeX_3^+$ in $TeCl_4$, $TeBr_4$ and $TeI_4$, $TeO_3^{2-}$ anions in tellurites, and bridged square-pyramidal $TeF_4$ all have a stereochemically active pair of non-bonding electrons with presumably greater 5*p*-character. These all show a more negative chemical isomer shift, and the predicted reduction in the lone-pair contribution to the *s*-electron density at the nucleus is compatible with $\delta R/R$ being positive. The tellurium(VI) compounds such as $Te(OH)_6$ and $TeO_4^{2-}$ show the most negative shifts because the 5*s*-orbital is now involved in covalent bonding, the nephelauxetic effect of which causes a further reduction in the *s*-electron density at the nucleus. Unfortunately, it is not possible to make reliable ab-initio molecular-orbital calculations for compounds in this area of the periodic table, and the relative occupations and hybridization of the various orbitals involved in bonding can only be estimated.

Neighbouring $^{121}Sb$ shows large chemical isomer shifts in the 37.1-keV resonance with a range of about 20 mm $s^{-1}$, and $\delta R/R$ is negative. The two common oxidation states Sb(III) and Sb(V) are comparatively distinct with the latter absorbing at more positive velocities. Some typical values of the Mössbauer parameters are collected in Table 4.3. The most negative chemical isomer shifts found in antimony are for the octahedral $[Sb^{III}Cl_6]^{3-}$ anions. Here

Table 4.3 $^{121}Sb$ Mössbauer parameters for antimony compounds

| Compound | | $\delta/(mm\ s^{-1})$ relative to $^{121}Sb$ in $SnO_2$ | $e^2qQ_g$ $/(mm\ s^{-1})$ |
|---|---|---|---|
| Antimony halides (at 80 K) [17] | | | |
| $K_3[SbCl_6]$ | | −17.8 | – |
| $Cs_3[SbCl_6]$ | | −17.7 | – |
| $[NH_4]_3[SbCl_6]$ | | −16.8 | – |
| $[Co(NH_3)_6][SbCl_6]$ | | −19.3 | – |
| $Rb_2[SbCl_6]$ | | −19.2<br>−2.4 | –<br>– |
| $Rb[SbCl_6]$ | | −2.3 | – |
| Organo-antimony(III) (at 4.2 K) [18] | | | |
| $Ph_3Sb$ | | −9.4 | +17 |
| Organo-antimony(V) (at 4.2 K) [18] | | | |
| $Ph_3SbF_2$ | | −4.69 | −22.0 |
| $Ph_3SbCl_2$ | | −6.02 | −20.6 |
| $Ph_3SbBr_2$ | | −6.32 | −19.8 |
| $Ph_3SbI_2$ | | −6.72 | −18.1 |
| $Ph_4SbF$ | | −4.56 | −7.2 |
| $Ph_4SbCl$ | | −5.26 | −6.0 |
| $Ph_4SbBr$ | | −5.52 | −6.8 |
| $Ph_4SbI$ | (at 80 K) | −5.6 | ? |
| $Ph_4SbNO_3$ | | −5.49 | −6.4 |
| $Ph_4SbClO_4$ | (at 80 K) | −5.9 | ? |

again the lone-pair has a high 5*s*-character and is sterically non-active [17], and with the negative sign for $\delta R/R$ results in the large negative shifts. The values are sensitive to the nature of the cation, there being a large difference between for example the $[NH_4]^+$ and $[Co(NH_3)_6]^{3+}$ salts. The caesium and potassium salts are both known to be distorted from octahedral symmetry in the solid state, and the evidence supports a small involvement of the 5*s*-orbital in the bonding scheme to cause a change in the *s*-electron density at the nucleus. No such cation effects were found in the isoelectronic tellurium compounds [16]. The compound $Rb_2SbCl_6$ shows two distinct absorptions [17] at $-19.2$ mm $s^{-1}$ from $Sb^{III}$ and at $-2.4$ mm $s^{-1}$ from $Sb^V$ and should be reformulated as

$Rb_4[Sb^{III}Sb^{V}Cl_{12}]$. This is a further example of the determination of formal oxidation states.

In compounds where there is a quadrupole splitting, the complex pattern from the $I_g = \frac{5}{2} \rightarrow I_e = \frac{7}{2}$ transition allows, in principle, the determination of both $e^2qQ$ and $\eta$, but in many cases the resolution is such that only a single asymmetric envelope is seen and only an approximate value of $e^2qQ$ can be obtained.

Data for some simple organo-antimony derivatives can be interpreted readily in terms of the covalent bonding [18]. Triphenylantimony, $Ph_3Sb$, is believed to be pyramidal with a stereochemically active lone-pair of electrons. It shows a substantial quadrupole splitting with $e^2qQ_g$ positive. Since $Q_g$ is known to be negative, the implication is that there is an excess of electron density in the *z*-direction (along the pyramidal axis), i.e. the lone-pair contains considerable $p_z$ character.

Triorgano-antimony(V) halides all have large negative quadrupole splittings (Table 4.3), indicative of excess *p*-electron density in the equatorial plane. The halogens are apical along the axis of the trigonal bipyramid in all positively characterized examples. As *p*-electron density is withdrawn along the *z*-axis, the quadrupole splitting increases in the order $Ph_3SbI_2 < Ph_3SbBr_2 < Ph_3SbCl_2 < Ph_3SbF_2$. At the same time *s*-electron density is also withdrawn to the axial ligands, and the chemical isomer shift becomes less negative. Thus the Sb-X bonds have both *p*- and *s*-character.

Compounds of the type $Ph_4SbX$ also show significant variation in parameters with change in X, and clearly favour covalent bonding of the halogen rather than a $Ph_4Sb^+$ cationic structure. A possible exception to this is the perchlorate which has the most negative shift. If they are all isostructural with $Ph_4SbOH$ which has an apical OH group, then the same interpretation already given to the $Ph_3SbX_2$ series can be applied.

## 4.4 Tin

Some of the difficulties encountered in using the $^{119}Sn$ resonance to study molecular compounds were referred to in Chapter 3. In particular, the failure of many unsymmetrical tin(IV) molecules to show a detectable quadrupole splitting of the $I_e = \frac{3}{2}$ excited state is both unexpected and still unexplained quantitatively. The confusion caused by uncertainties in the stereochemistry of many organotin compounds, and our poor knowledge regarding the origins of the electric field gradient at the tin nucleus have resulted in the extensive use of empiricism. However, the problems are no greater than with any of the other spectroscopic techniques, and it is gratifying to observe the way in which Mössbauer spectroscopy has drawn attention to many fundamental issues in the understanding of the chemistry of tin, which in turn it is helping to solve.

Compounds in the tin(II) oxidation state are known to feature, with few exceptions, a pyramidal bonding arrangement of three ligands about the tin. Early theories regarding the origin of the electric field gradient predicted that the sign of $e^2qQ$ in these compounds could be either positive or negative. However, recent measurements on $SnF_2$, SnO, SnS, $Sn_3(PO_4)_2$, $SnC_2O_4$, $NaSnF_3$, $NaSn_2F_5$, $SnSO_4$, $Sn(HCO_2)_2$, $Sn(CH_3CO_2)_2$ and $K_2Sn(C_2O_4)_2.H_2O$ in applied magnetic fields [19, 20] have shown that $e^2qQ$ is positive in every case. As $Q$ is known to be negative, the sign of $e^2qQ$ is consistent with an excess of electron density in the $p_z$ orbital (for which $q$ is negative, see p.35). Thus the lone-pair has a significant $p_z$ character which becomes the dominant contribution to the electric field gradient because of tighter binding to the nucleus. However, neither simple covalent nor point-charge arguments have been able to predict successfully the magnitude of the gradient.

A salutory warning as to the dangers inherent in the interpretation of small differences between related compounds is provided by data for the trihalogeno complexes, $[SnXYZ]^-$ where X,Y,Z = Cl, Br, I. The chemical isomer shifts for the $[Bu^n_4N]^+$ salts of these compounds are correlated in Fig.4.4 with the average Mulliken electronegativity of the ligands, $\chi_M$, and a linear relationship is found [21]. A similar correlation exists between the chemical isomer shift and the quadrupole splitting. The electric field gradient is influenced by the polarity of the tin-ligand bonds and increases with increasing polarization, presumably because the $p_z$ character of the lone-pair becomes in greater imbalance with respect to the covalent bonding orbitals. The trend in quadrupole splitting values is $\Delta(SnI_3^-) = 0.79$ mm s$^{-1}$ $< \Delta(SnBr_3^-) = 1.02$ mm s$^{-1}$ $< \Delta(SnCl_3^-) = 1.13$ mm s$^{-1}$. This does not necessarily imply that the $p_z$ character has increased.

An unexpected and startling observation is the large difference in both Mössbauer parameters between the $[Bu^n_4N]^+$ salts and cor-

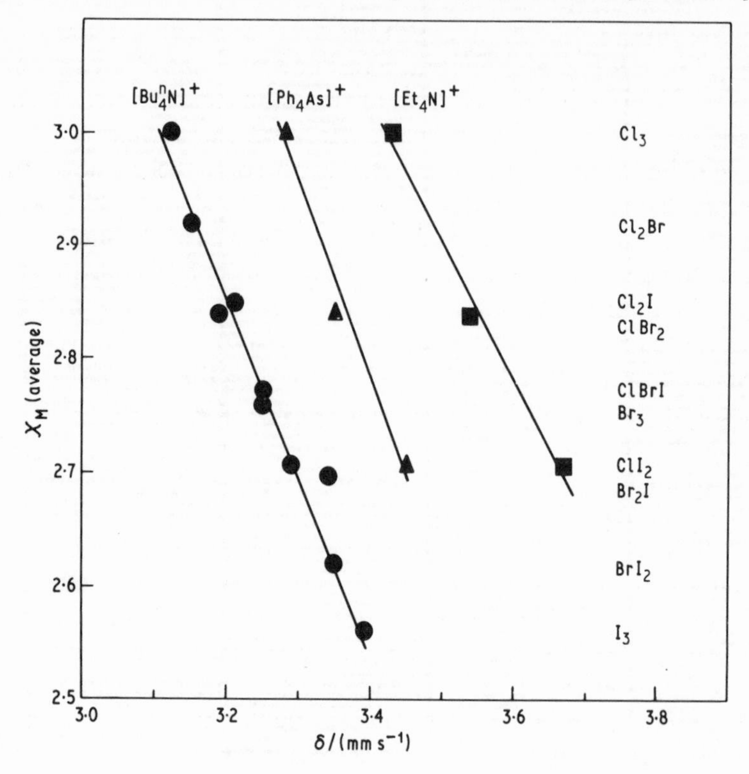


Fig. 4.4 A correlation of the chemical isomer shift in tin(II) compounds of the type $[SnXYZ]^-$, where X, Y, Z = Cl, Br, I, with the average Mulliken electronegativity of the ligands, $\chi_M$. Note the dependence on the nature of the cation.

responding $[Ph_4As]^+$ and $[Et_4N]^+$ salts. In fact the chemical isomer shift difference between $[Bu^n_4N]SnCl_3$ and $[Et_4N]SnCl_3$ is greater than between $[Bu^n_4N]SnCl_3$ and $[Bu^n_4N]SnI_3$. While the vibrational spectra of the $[Bu^n_4N]^+$ salts in solution are essentially the same as in the solid, this is not the case with the $[Et_4N]^+$ salts. The $[Ph_4As]^+$ salts appear to be intermediate. One can only conclude that there is a substantial solid-state interaction within the $[Et_4N]^+$ salts, although whether this is between the anion and cation, or between two anions has not been ascertained. It seems possible that many difficulties experienced in rationalization of data for tin(II) compounds are a result of nominally weak intermolecular interactions having a disproportionately large influence on the lone-pair of electrons.

A chemical isomer shift-electronegativity correlation has also been shown for tin(IV) compounds. The analogous $[SnX_4Y_2]^{2-}$ anions (X, Y = Cl, Br, I) also give a linear correlation [22, 23]. Although any quadrupole splitting in the unsymmetrical $[SnX_4Y_2]^{2-}$ species is unresolved, a small degree of line broadening is seen which is proportional to the electronegativity difference in X and Y. This implies that a small electric field gradient is in fact present. The corresponding fluorides appear to be anomalous. However, it is interesting to note that the $(ClO_2)_2SnF_6$ salt has a large splitting of 1.01 mm $s^{-1}$ as well as an unusually positive chemical isomer shift for an $SnF_6^{2-}$ salt [24]. A more general spectroscopic study of several fluoride salts has indicated that anion-cation interactions are the rule rather than the exception.

The tetrahalides $SnX_4$ (X = Cl, Br, I) in which the tin is tetrahedrally co-ordinated also show a linear correlation of the chemical isomer shift with electronegativity, as do the oxyhalides $SnOX_2$ (X = Cl, Br, I). The slope of the line is different for each series, and in the case of the oxyhalides is of opposite sign. The behaviour observed is strongly dependent on stereochemistry, and results from a subtle balance between the 5*p*- and 5*s*-character in the various bonds and the shielding effect of 5*p*-electrons on the 5*s*-electrons. The importance of 5*d*-character in this context cannot be assessed at the present time, other than that the effect on the quadrupole splitting would seem to be less important. For any series of compounds where the co-ordination and geometry are the same for all members, and one of the ligands is systematically varied, a gradation of properties can be expected which allows the bonding characteristics of unfamiliar ligands to be assessed in relation to the series.

The factors governing the Mössbauer parameters of organotin compounds can be particularly obscure. For example in any series $R_xSnX_{4-x}$ ($x$ = 0–4, R = alkyl or aryl, X = halogen) the chemical isomer shift and quadrupole splitting are both considerably greater when $x$ = 2 or 3. There is no obvious correlation between the shift and the nominal degree of substitution as one would expect if the major factor were the electron withdrawing power of the groups bonded to tin. Thus in the series $Me_4Sn$, $Me_3SnBr$, $Me_2SnBr_2$, $MeSnBr_3$ and $SnBr_4$ the values for the chemical isomer shift and quadrupole splitting are $\delta$ = 1.21, 1.38, 1.59, 1.41, 1.15 mm $s^{-1}$ and $\Delta$ = 0, 3.20, 3.41, 1.91, 0 mm $s^{-1}$ respectively.

It was shown in Chapter 3 that the magnitude of the quadrupole splitting in tin(IV) compounds does show a certain degree of correlation with the stereochemistry. Nevertheless, a direct determination of the sign of $e^2qQ$ is of great assistance in determining the origin of the *p*-electron imbalance, and enables an assignment of the co-ordination number and stereochemistry in many cases (see Chapter 3 for a fuller discussion). Although the use of an externally applied magnetic field to determine the sign is a comparatively recent development, such data as are available show convincingly that the electric field gradient originates primarily from asymmetric occupation of the $\sigma$-bonding orbitals, rather than from $\pi$-bonding to bridging groups or indirectly from lattice effects. The relative importance of *p*- and *d*-electrons in $\sigma$-bonding remains unknown, although it may be presumed that the contribution of *d*-electrons to the electric field gradient will be comparatively small.

The effects of distortion from regular geometry are probably greater than has been generally assumed. A good example is given by $Ph_2SnCl_2$ which has a distorted tetrahedral geometry in the solid state [25]. The mean bond angles are Cl-Sn-Cl 100°, Cl-Sn-C 107° and C-Sn-C 125.5° with a mean Sn-Cl bond distance of 2.346 Å (234.6 pm). The third-nearest chlorine neighbour is at 3.77 Å, so that the structure cannot be described as polymeric. The $^{119}Sn$ quadrupole splitting in the pure solid is 2.80 mm $s^{-1}$, but a different value is found in crystalline and glassy matrices made by freezing solutions of $Ph_2SnCl_2$ [26]. In a non-basic solvent the quadrupole splitting decreases by about 0.2 mm $s^{-1}$ to $\sim$2.61 mm $s^{-1}$ and it is thought that this value is more representative of an isolated unrestrained molecule. Basic solvents give an increased splitting of $\sim$3.5 mm $s^{-1}$. Presumably in this case two solvent molecules co-ordinate to the $Ph_2SnCl_2$ to give a distorted *trans*-octahedral complex with phenyl groups in the axial positions.

The anomalous sign of $e^2qQ$ in *cis*-$Me_2Sn(oxin)_2$ referred to in Chapter 3 is undoubtedly a result of the large distortion from pseudo-regular geometry (the C-Sn-C bond angle is 111°). In the case of $Me_2SnCl_2$ the stereochemistry is effectively a badly distorted tetrahedron, but the crystal packing is such that it approximates to a polymeric chain with *trans*-methyl groups and bridging chlorine atoms [27]. It therefore seems likely that many of the apparent anomalies in the magnitude of the quadrupole splittings for series of related compounds can be directly attributed to distortions from pseudo-regular tetrahedral, trigonal bipyramidal or octahedral stereochemistry.

One of the few tin compounds where a more quantitative description of the bonding has been possible is $[(\pi\text{-}C_5H_5)Fe(CO)_2]_2SnCl_2$. The X-ray structure is known, and the geometry of the $Fe_2SnCl_2$ unit has $C_2$ symmetry. The $^{119}Sn$ spectrum in an applied magnetic field is shown in Fig.4.5, together with computer simulations for various values of the asymmetry parameter $\eta$ [28], and it can be seen that $\eta$ is approximately 0.65 with $e^2qQ$ positive. It is possible to represent the contribution of each ligand to the electric field gradient using a point charge approximation in which the effective value of $eq$ for the Sn-Fe bond is represented as [Fe] etc. The molecular geometry is shown in Fig.4.6. The Fe-Sn-Fe plane is perpendicular to the Cl-Sn-Cl plane, and therefore the three

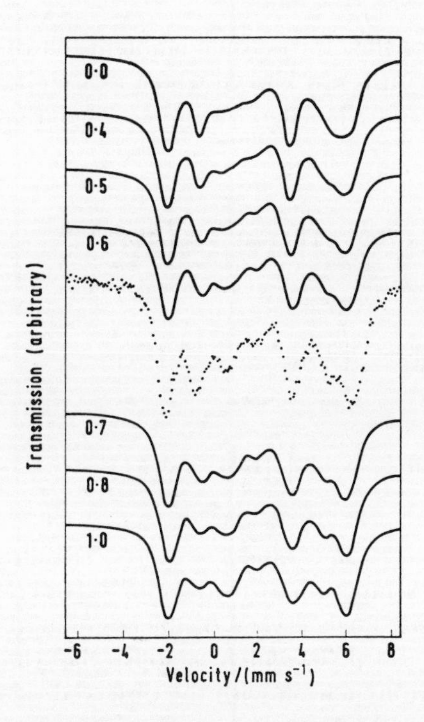


Fig. 4.5 The Mössbauer spectrum of $[(\pi\text{-}C_5H_5)Fe(CO)_2]_2SnCl_2$ at 4.2 K with a magnetic field of flux density 5 T applied perpendicular to the direction of observation. The solid curves are computed predictions for different values of $\eta$. ([28], Fig. 1).

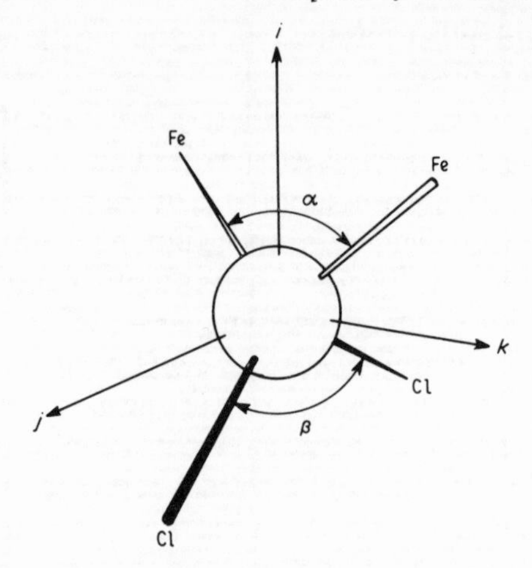


Fig. 4.6 The molecular geometry of $[(\pi\text{-}C_5H_5)Fe(CO)_2]_2SnCl_2$.

principal axes of the electric field gradient tensor lie in these planes. *A priori* it is not known which axis is the principal $z$ axis, so the three axes are labelled $i$, $j$ and $k$. Then the relationships can be derived

$$V_{ii} = 2\left(3\cos^2\frac{\alpha}{2} - 1\right)[\mathrm{Fe}] + 2\left(3\cos^2\frac{\beta}{2} - 1\right)[\mathrm{Cl}]$$

$$V_{jj} = -2[\mathrm{Fe}] + 2\left(3\sin^2\frac{\beta}{2} - 1\right)[\mathrm{Cl}]$$

$$V_{kk} = 2\left(3\sin^2\frac{\alpha}{2} - 1\right)[\mathrm{Fe}] - 2[\mathrm{Cl}]$$

The angles are $\alpha = 128.6°$ and $\beta = 94.1°$. The direction and sign of $V_{zz}$ is a complex function of the ratio $R = [\mathrm{Fe}]/[\mathrm{Cl}]$. The quadrupole splitting in a single crystal of the compound shows unequal line intensities (see p.40) which establish that the principal axis, $z$, is along $k$. In combination with the observed value for the asymmetry parameter this leads to a unique value for $R$ of 1.2. It may therefore be concluded that the more electronegative halogen atoms withdraw a greater proportion of $5p$-electron density from the tin atom in $\sigma$-bonding, and therefore make a smaller contribution than the iron to the electric field gradient.

This compound is also of interest in being one of several for which the crystal structure shows unusually short Sn-Fe and long Sn-Cl bond distances. It was originally thought that this was the result of intermolecular interactions or crystal-packing effects, but the similarity of the Mössbauer spectra of the pure solid with those of frozen solutions in inert matrices confirms that the structure reflects the internal bonding of the molecule itself.

## 4.5 Covalent iron compounds

The transition metals differ from the main-group elements in that the valence *d*-orbitals are only partially involved in chemical bonding, and therefore retain a degree of identity which is particularly useful. Many compounds have one or more unpaired electrons, and being of essentially *d*-character, these cause characteristic effects which are considered separately in the next chapter. Where the covalency is strong, this proves to have a substantial influence on the Mössbauer spectrum.

In considering covalent effects in more detail, it is convenient to consider compounds of iron(II) in the low-spin ($S = 0$) configuration where paramagnetic effects are absent. Nevertheless, the discussion is directly applicable to for example the chemical isomer shift in iron(III) low-spin ($S = \frac{1}{2}$) complexes. A typical compound is potassium hexacyanoferrate, $K_4[Fe(CN)_6].3H_2O$, which has octahedral co-ordination in the $[Fe(CN)_6]^{4-}$ anion. The chemical isomer shift of $-0.04$ mm s$^{-1}$ (with respect to iron metal) at room temperature contrasts strongly with the typical values of +1.0 to +1.3 mm s$^{-1}$ for octahedral high-spin iron(II) compounds ($S = 2$). An octahedral ligand-field splits the five-fold degeneracy of the 3*d*-electrons into an upper doublet ($e_g$) and a lower triplet ($t_{2g}$) level. In the $S = 2$ compounds, the electron configuration is $t_{2g}^4 e_g^2$, but for $S = 0$ it becomes $t_{2g}^6$, i.e. a fully occupied triplet level. In discussing the cyanide complexes, it is particularly important to consider the effect of $\pi$-bonding between the $t_{2g}$ levels and the $\pi$-bonding and $\pi^*$-antibonding orbitals of the $CN^-$ ligands. With this in mind, the chemical isomer shift observed will be compounded from several factors:

(1) a direct contribution from 4*s*-electrons, and

(2) indirect contributions from $3d$-electrons by shielding the $s$-electrons, which may be considered in terms of:

(a) the purely ionic or non-bonding effect of a $3d^n$ configuration.

(b) covalency with the ligands in which filled ligand orbitals donate an effective total of $n_2$ electrons to the $3d$-orbitals.

(c) covalency with the ligands in which the metal $3d$-orbitals donate an effective total of $n_3$ $3d$-electrons to the $\pi^*$ ligand orbitals.

Thus the effective number of $3d$-electrons is

$$n_{\text{eff}} = n + n_2 - n_3$$

In molecular orbital terms, one can say that the metal $t_{2g}$ ($3d$) levels and the $\pi$ and $\pi^*$ ligand orbitals all have $t_{2g}$ symmetry and combine to form molecular orbitals with the same symmetry. The $e_g$ ($3d$) orbitals have the same symmetry as ligand $\sigma$-orbitals, so that a large part of $n_2$ will be $\sigma$-donation from the ligands.

Calculation of the relative importance of these effects is difficult, but one interesting example concerns a comparison of $[Fe(CN)_6]^{4-}$ and $[Fe(CN)_6]^{3-}$ made by Shulman and Sugano [29]. They estimated many of the molecular orbital parameters from electron spin resonance (E.S.R.) data, and found that the degree of $\pi$ back-donation as typified by $n_3$ is about one electron greater in $[Fe(CN)_6]^{4-}$ than in $[Fe(CN)_6]^{3-}$, while the term $n_2$ remains about the same. Therefore the value of $n_{\text{eff}}$ is similar in both compounds, and explains why they have almost the same chemical isomer shift. Although the hexacyanoferrate(II) nominally has an extra $3d$-electron, its effect is largely nullified by an additional delocalization of the $t_{2g}$ orbitals to the ligands. However, it is important to realize that the two configurations are distinct. The full molecular orbital schemes still feature an unpaired-electron in the hexacyanoferrate(III), whereas the hexacyanoferrate(II) is diamagnetic.

The Mössbauer spectrum of a hexacyanoferrate(II) is only a single line because of the regular octahedral geometry, but the same is not true of substituted cyanides where a quadrupole splitting will be seen. A simple example is the so-called nitroprusside ion, $[Fe(CN)_5(NO)]^{2-}$, which shows a splitting of 1.705 mm s$^{-1}$ [30]. The ground state configuration for the axial $C_{4v}$ symmetry is

$(d_{xz}, d_{yz})^4 (d_{xy})^2$. The $(d_{xz}, d_{yz})$ orbitals are strongly delocalized by $\pi$-donation to the $2p$-orbitals of the nitrosyl group, and there is also some $\pi$-donation from the $d_{xy}$ orbital to the equatorial cyanides.

The electric field gradient may be presumed to originate mainly from the unequal contributions of these $3d$-orbitals. If, as with the hexacyanoferrate(II), we define effective occupation numbers $n_{xy}$ etc., it becomes possible to write the value of $q$ in the quadrupole splitting $e^2qQ$ in terms of the value for a single $d$-electron of $(4/7)(1-R)\langle r^{-3}\rangle/(4\pi\epsilon_0)$ as

$$q = \frac{(1-R)}{4\pi\epsilon_0}\langle r^{-3}\rangle\left[+\frac{4}{7}n_{xy} - \frac{2}{7}(n_{xz} + n_{yz})\right]$$

The value of $n_{xy}$ (incorporating the delocalization to the cyanides) can be estimated to be approximately 1.74 via the E.S.R. orbital reduction factor in $[Fe(CN)_6]^{3-}$. Using an estimated value for the quadrupole splitting from a single $d$-electron of $\sim$4.0 mm s$^{-1}$, and the experimental value of 1.705 mm s$^{-1}$, it is possible to derive $(n_{xz} + n_{yz}) \simeq 2.63$. If the delocalization to the nitrosyl is indeed greater than to the cyanide, then this figure for the occupation of $d_{xz}$ and $d_{yz}$ can be used to derive an approximate molecular orbital

$$\psi_{xz} = 0.81\, d_{xz} + 0.58\, \pi^* \text{(NO)}$$

The resultant occupation figure for the $\pi^*$ (NO) orbital of 34% is an upper limit, and is in good agreement with independent estimates of about 25%.

The nitroprusside ion has an unusually low chemical isomer shift for an iron(II) cyanide complex ($-0.258$ mm s$^{-1}$ relative to iron metal at 298 K). The effective $3d$-population, $n_{\text{eff}}$, is only $\sim$4.4, and there is an unusually high $4s$-population of $\sim$0.5. These two factors combine to produce the exceptionally low shift observed.

The related complex ion $[Fe(CN)_5(NO)]^{3-}$ has an extra electron in an orbital of essentially $d_{z^2}$ character, and in frozen solutions its parameters are a quadrupole splitting of $\Delta = +1.90$ mm s$^{-1}$ and a chemical isomer shift of $\delta = +0.1$ mm s$^{-1}$ [31]. In this case the value of $q$ is given by

$$q = \frac{(1-R)}{4\pi\epsilon_0}\langle r^{-3}\rangle[+\tfrac{4}{7}n_{xy} - \tfrac{2}{7}(n_{xz}+n_{yz}) - \tfrac{4}{7}n_{z^2}]$$

One would expect the value of $n_{xy}$ to be little changed from that in the nitroprusside, and consideration of E.S.R. data leads to an estimate for $n_{z^2}$ of 0.6. The experimental value for the quadrupole splitting can then be used to derive the values $(n_{xz}+n_{yz}) \simeq 1.6$, $n_{xy} \simeq 1.9$, $n_{z^2} \simeq 0.6$, giving $n_{\text{eff}} \simeq 4.1$. Although these values are approximate, the large decrease in $(n_{xz}+n_{yz})$ compared to the nitroprusside reflects the large increase in $\pi$-donation to the $NO^+$ ligand as a result of a formal reduction from Fe(II) to Fe(I). This offsets the increase in electron density about the central atom associated with the addition of a non-bonding electron.

It will now be apparent that it is difficult to estimate the relative effects of $\sigma$- and $\pi$-bonding from the comparatively small amount of information obtainable from a single compound. A more fruitful approach is to compare data for large numbers of related complexes. In Chapter 3 it was shown that *cis*- and *trans*-isomers of low-spin iron(II) compounds may be distinguished by a 1:2 ratio of the quadrupole splittings. From data for large numbers of low-spin iron(II) compounds, the chemical isomer shift has been found to be an approximately additive property of the number and type of the ligands [32, 33]. Any contributions to the shift from either the second-order Doppler shift or zero-point motion are ignored on the grounds that for comparative purposes they are self-cancelling in similar compounds. Similarly, the quadrupole splitting, $\Delta$, can be expressed in terms of partial quadrupole splittings such that for the three stereochemistries $FeAB_5$, *cis*-$FeA_2B_4$ and *trans*-$FeA_2B_4$

$$\Delta_{AB_5} = 2(\text{PQS})_A = 2(\text{PQS})_B$$

$$\Delta_{cis} = -2(\text{PQS})_A + 2(\text{PQS})_B$$

$$\Delta_{trans} = 4(\text{PQS})_A - 4(\text{PQS})_B$$

The justification for these equations can be seen by considering the Appendix at the end of Chapter 3. The partial chemical isomer shift (PCS) and partial quadrupole splitting (PQS) values for a number of simple monodentate ligands are given in Table 4.4.

Table 4.4 Partial chemical isomer shift (PCS) and partial quadrupole splitting (PQS) data for low-spin 6-coordinate iron(II) compounds

| Ligand | PCS/(mm $s^{-1}$)[a] | PQS/(mm $s^{-1}$)[b] |
|---|---|---|
| $NO^+$ | −0.20 | +0.02 |
| $H^-$ | −0.08 | −1.04 |
| RNC | 0.00 | −0.69 |
| $CN^-$ | 0.01 | −0.84 |
| $SnCl_3^-$ | 0.04 | −0.43 |
| $NCS^-$ | 0.05 | −0.49 |
| $NCO^-$ | 0.06 | −0.50 |
| $NH_3$ | 0.07 | −0.51 |
| $N_3^-$ | 0.08 | −0.38 |
| $H_2O$ | 0.10 | −0.44 |
| $Cl^-$ | 0.10 | −0.30 |
| $Br^-$ | 0.13 | −0.28 |
| $I^-$ | 0.13 | −0.29 |

(a) based on stainless steel at 295 K as reference. Calculated values for $\delta$ can be converted to iron metal as standard by subtracting 0.09 mm $s^{-1}$.
(b) based on −0.30 mm $s^{-1}$ for $Cl^-$ as an arbitrary value.

The compound *cis*-$Fe(SnCl_3)_2(RNC)_4$ would be expected to show $\Delta = -2 \times (-0.43) + 2 \times (-0.69) = -0.52$ mm $s^{-1}$ (observed 0.50) and $\delta = 2 \times (0.04) + 2 \times (0.00) = +0.08$ mm $s^{-1}$ (observed +0.11) relative to the reference point of $^{57}Co$ in stainless steel. The partial shift and splitting values can be related to the bonding characteristics of the ligands. This can be illustrated by considering cyanides of the type $[Fe(CN)_5L]^{n-}$ where L is a variable ligand. For instance the chemical isomer shift increases in the order $NO^+ < CO < CN^- < Ph_3P < SO_3^{2-} < NO_2^- \simeq NH_3 \simeq Ph_3As \simeq Ph_3Sb$, corresponding to a decrease in the $\pi$ back-donation to the ligand antibonding orbitals.

The partial chemical isomer shift value will decrease with increasing $\sigma$-bonding from the ligand to hybrid $d^2sp^3$ orbitals on the metal. The direct increase in 4*s*-density is greater than the increased shielding due to 3*d*-augmentation, and $\delta R/R$ is negative. An increase in $\pi$ back-bonding to the ligand from the $t_{2g}$ (3*d*) orbitals decreases the shielding of the inner *s*-orbitals and also causes a decrease in chemical isomer shift. The partial quadrupole splitting will become more negative with increased $\sigma$-bonding and more positive with increased $\pi$ back-bonding.

$$PCS = -k(\sigma + \pi)$$

$$PQS = +K(\pi - \sigma)$$

where k and K are constants of proportionality. Thus the best $\sigma$-donor ($H^-$) and the best $\pi$-acceptor ($NO^+$) give the most negative PCS values, but are at opposite extremes of the PQS scale.

This type of treatment is very useful in assessing the relative importance of $\sigma$- and $\pi$-bonding for a series of ligands, but cannot be applied to series of compounds which differ in co-ordination or geometry, or show non-additive effects in the bonding such as a strong 'trans-effect'. For this reason the covalent iron carbonyl derivatives do not readily lend themselves to this kind of interpretation, although where series of related compounds with similar stereochemistry can be found, it is still possible to make empirical correlations with the $\sigma$- and $\pi$-bonding characteristics of particular ligands.

One particularly interesting organometallic derivative is $(\pi\text{-}C_5H_5)_2Fe$, bis(*pentahapto*-cyclopentadienyl)iron(II) commonly known as ferrocene, which gives a large temperature-independent quadrupole splitting of 2.37 mm $s^{-1}$. The corresponding cation, $(\pi\text{-}C_5H_5)_2Fe^+$, contrasts in that splitting is almost absent. This was suspected and later confirmed by Collins [34] to be the result of a chance cancellation of the various contributions to the electric field gradient tensor. Several types of theoretical bonding scheme are available for the ferrocene molecule. A crystal field calculation by Matsen predicted a negative sign for $e^2qQ$, whereas molecular orbital schemes by Ballhausen (using Watson's wavefunctions for iron) and by Shustorovich and Lyatkina (using Slater wavefunctions) both predict $e^2qQ$ to be positive. Collins determined the sign experimentally by the first application of the magnetic perturbation method to be reported in the literature, and found that it was positive.

The molecular orbital scheme, adopted from Dahl and Ballhausen, is shown in Table 4.5. The molecular orbital wavefunctions ($\psi$) are taken as linear combinations of metal ($\mu$) and ligand ($\rho$) orbitals. The effective occupation of the metal atomic orbitals is then given by the squares of the metal coefficients. The 4*s*-orbitals do not contribute to the electric field gradient tensor, and there is good evidence to show that any contribution from the 4*p*-orbitals

Table 4.5 Molecular-orbital calculations for ferrocene

| Wave function (Increasing energy ↑) | No. of electrons | $\left(\text{metal coeff.}\right)^2$ | net Δ* $\|4p_0>$ | net Δ* $\|3d_0>$ |
|---|---|---|---|---|
| $\psi(e_{2g}) = 0.898\mu(3d_2) + 0.440\rho(e_{2g}^+)$ | 4 | 0.8064 | | −3.222 |
| $\psi(e_{1g}) = 0.454\mu(3d_1) + 0.891\rho(e_{1g}^+)$ | 4 | 0.2061 | | +0.411 |
| $\psi(a_{1g}) = \mu(3d_0)$ | 2 | 1.000 | | +2.000 |
| $\psi(e_{1u}) = 0.591\mu(4p_1) + 0.807\rho(e_{1u}^+)$ | 4 | 0.3493 | −0.693 | |
| $\psi(a_{2u}) = 0.471\mu(4p_0) + 0.882\rho(a_{2u})$ | 2 | 0.2218 | +0.444 | |
| $\psi(a_{1g}) = 0.633\mu(4s) + 0.774\rho(a_{1g})$ | 2 | 0.4007 | | |
| | | | −0.249 | −0.811 |

$(\pi - C_5H_5)_2Fe$ $\quad e^2qQ/2 = -0.249|4p_0> - 0.811|3d_0>$

$(\pi - C_5H_5)_2Fe^+$ $\quad e^2qQ/2 = -0.249|4p_0> - 0.005|3d_0>$

* The contributions to the quadrupole splitting in units of that from a $|4p_0>$ or $|3d_0>$ electron. Both quantities are defined as negative.

will be small in comparison to that from the 3*d*, because of the greater radial expansion of the former. Each individual orbital generates a contribution to the field gradient and to $e^2qQ$ along the symmetry axis of the molecule, and in the last two columns of the Table these are normalized in terms of that from the $|4p_0>$ function and the $|3d_0>$ function (which will be of the order of −3.6 mm $s^{-1}$ per electron). We can now see that in ferrocene one predicts a quadrupole splitting of $-0.811|3d_0> = \sim+2.9$ mm $s^{-1}$. Removal of an electron from the $\psi(e_{2g})$ orbitals (to generate the cation) will effectively remove 0.8064 of a $|3d_0>$ electron and reduce the splitting to $-0.005|3d_0> = \sim+0.2$ mm $s^{-1}$. This is in excellent agreement with observation.

The same principles are of course applicable to the paramagnetic compounds, but the metal-ligand interaction in for example a high-spin $Fe^{2+}$ compound is much weaker than in a low-spin Fe(II) compound. The observed $^{57}Fe$ chemical isomer shifts at 4.2 K (in mm $s^{-1}$ with respect to Fe metal) in for example the anhydrous iron(II) halides are $FeF_2$ (1.48), $FeCl_2$ (1.16), $FeBr_2$ (1.12),

$FeI_2$ (1.04). These values correspond to an increase in *s*-electron density due to reduced 3*d*-shielding and an increase in covalency along the series. This is confirmed by the crystal structures which change from the rutile type for $FeF_2$ to the $CdCl_2$-type for $FeCl_2$ and the $CdI_2$-type for $FeBr_2$ and $FeI_2$. All feature a distorted 6-coordination to the halogen. In a tetrahedral ion such as $[FeCl_4]^{2-}$ however, the shift of 0.90 mm $s^{-1}$ is lower than in an octahedral co-ordination because of a substantial increase in covalency. This effect is also found in iron oxides and sulphides.

The influence of covalency in paramagnetic compounds has not been studied systematically. Perhaps this is because of the greater interest which attaches to phenomena associated with the unpaired electron. These are considered in detail in the following chapter.

## References


[1] Watson, R. E. (1960) *Phys. Rev.*, **118**, 1036.

[2] Watson, R. E. (1960) *Phys. Rev.*, **119**, 1934.

[3] Walker, L. R., Wertheim, G. K. and Jaccarino, V. (1961) *Phys. Rev. Letters*, **6**, 98.

[4] Duff, K. J. (1974) *Phys. Rev. B*, **9**, 66.

[5] Hafemeister, D. W., Pasquali, G. de and Waard, H. de (1964) *Phys Rev.*, **135**, B1089.

[6] Perlow, G. J. and Perlow, M. R. (1966) *J. Chem. Phys.*, **45**, 2193.

[7] Ehrlich, B. S. and Kaplan, M. (1969) *J. Chem. Phys.*, **50**, 2041.

[8] Das, T. P. and Hahn, E. L. (1958) *Nuclear Quadrupole Resonance Spectroscopy*, Supplement I of Solid State Physics, Academic Press Inc., New York.

[9] Pasternak, M., Simopoulos, A. and Hazony, Y. (1965) *Phys. Rev.*, **140**, A1892.

[10] Bukshpan, S., Goldstein, C. and Sonnino, R. (1968) *J. Chem. Phys.*, **49**, 5477.

[11] Pasternak, M. and Sonnino, T. (1968) *J. Chem. Phys.*, **48**, 1997.

[12] Pasternak, M. and Sonnino, T. (1968) *J. Chem. Phys.*, **48**, 2009.

[13] Bukshpan, S. and Sonnino, T. (1968) *J. Chem. Phys.*, **48**, 4442.

[14] Bukshpan, S. and Herber, R. H. (1967) *J. Chem. Phys.*, **46**, 3375.

[15] Bukshpan, S. (1968) *J. Chem. Phys.*, **48**, 4242.

[16] Gibb, T. C., Greatrex, R., Greenwood, N. N. and Sarma, A. C. (1970) *J. Chem. Soc. (A)*, 212.

[17] Birchall, T., Della Valle, B., Martineau, E. and Milne, J. B. (1971) *J. Chem. Soc. (A)*, 1855.

[18] Long, G. G., Stevens, J. G., Tullbone, R. J. and Bowen, L. H. (1970) *J. Amer. Chem. Soc.*, **92**, 4230.

[19] Gibb, T. C., Goodman, B. A. and Greenwood, N. N. (1970) *Chem. Comm.*, p.774.
[20] Donaldson, J. D., Filmore, E. J. and Tricker, M. J. (1971) *J. Chem. Soc. (A)*, 1109.
[21] Goldstein, M. and Tok, G. C. (1971) *J. Chem. Soc. (A)*, 2303.
[22] Herber, R. H. and Cheng, Hwan Sheng (1969) *Inorg. Chem.*, **8**, 2145.
[23] Clausen, C. A. and Good, M. L. (1970) *Inorg. Chem.*, **9**, 817.
[24] Carter, H. A., Qureshi, A. M., Sams, J. R. and Aubke, F. (1970) *Canad. J. Chem.*, **48**, 2853.
[25] Greene, P. T. and Bryan, R. F. (1971) *J. Chem. Soc. (A)*, p.2549.
[26] Herber, R. H. (1973) *J. Inorg. Nucl. Chem.*, **35**, 67.
[27] Davies, A. G., Milledge, H. J., Puxley, D. C. and Smith, P. J. (1970) *J. Chem. Soc. (A)*, 2862.
[28] Gibb, T. C., Greatrex, R. and Greenwood, N. N. (1972) *J. Chem. Soc. (A)*, p.238.
[29] Shulman, R. G. and Sugano, S. (1965) *J. Chem. Phys.*, **42**, 39.
[30] Danon, J. and Iannarella, L. (1967) *J. Chem. Phys.*, **47**, 382.
[31] Raynor, J. B. (1971) *J. Inorg. Nucl. Chem.*, **33**, 735.
[32] Bancroft, G. M., Mays, M. J. and Prater, B. E. (1970) *J. Chem. Soc. (A)*, p.956.
[33] Bancroft, G. M., Garrod, R. E. B. and Maddock, A. G. (1971) *J. Chem. Soc. (A)*, 3165.
[34] Collins, R. L. (1965) *J. Chem. Phys.*, **42**, 1072.

CHAPTER FIVE

# *Electronic Structure and Bonding: Paramagnetic Compounds*

In Chapter 4 it was shown how the chemical isomer shift and quadrupole splitting in diamagnetic compounds are dependent on the oxidation state and covalent bonding of the resonant atom. The chemical isomer shift in paramagnetic compounds is governed by the same principles, and it was therefore convenient to consider the influence of covalency in high-spin $Fe^{2+}$ and low-spin Fe(III) compounds in Section 4.5. However, the presence of an unpaired electron can result in magnetic and quadrupole interactions which are intimately related to the orbital state of the atom. It is therefore appropriate to consider these separately. The major part of this chapter is concerned with iron chemistry, but the ideas expressed are also applicable to other elements, and some of these are discussed separately at the end of the chapter.

## 5.1 Quadrupole interactions

The quadrupole interaction in a diamagnetic compound is determined mainly by asymmetry in the covalent bonding of the resonant atom. As a rule, the first excited electronic level is not accessible by thermal excitation (being at a much higher energy than $kT$ at room temperature where $k$ is Boltzmann's constant and $T$ is the temperature). The electronic configuration and the orbital occupation are therefore not sensitive to change in temperature, and in consequence the quadrupole splitting is almost independent of temperature. This behaviour is found in for example tin and low-spin iron(II) compounds. However, in many paramagnetic ions there is a substantial degree of thermal excitation to electronic levels which are only of the order of $kT$ above the ground state. The resulting temperature dependence of the orbital occupation produces characteristic effects on the quadrupole splitting.

The most familiar example of this is the high-spin $Fe^{2+}$ ion. The free-ion configuration $^5D$ ($t_{2g}^4 e_g^2$) signifies a single 3*d*-electron outside a spherical half-filled shell. In a ligand field of cubic symmetry such as regular octahedral, the $t_{2g}$ levels remain degenerate (as do the $e_g$) and there is no finite electric field gradient. However, if the symmetry is lowered to trigonal or tetragonal, further degeneracy is removed. The sixth electron is in the appropriate lowest-lying unfilled state, and now generates an electric field gradient at the iron nucleus. A ligand field of axial symmetry (i.e. a tetragonal distortion of the octahedron) splits the $t_{2g}$ state into a lower $d_{xy}$ singlet and an upper $d_{xz,yz}$ state. The sixth electron will be in the $d_{xy}$ level and (from Table 2.1) will generate a quadrupole splitting in proportion to $V_{zz}/e = q = +\frac{4}{7}\langle r^{-3}\rangle(1 - R)/(4\pi\epsilon_0)$. If the doublet lies lowest, then the electron is in $d_{xz,yz}$ and gives $q = -\frac{2}{7}\langle r^{-3}\rangle(1 - R)/(4\pi\epsilon_0)$. If the distortion is trigonal, i.e. an elongation along the (111) axis, the singlet ground state corresponds to $d_{z^2}$ and gives $q = -\frac{4}{7}\langle r^{-3}\rangle(1 - R)/(4\pi\epsilon_0)$.

A singlet $d_{xy}$ ground state is found for example in $[Fe(H_2O)_6](NH_4SO_4)_2$ for which $e^2qQ$ is positive, and a singlet $d_{z^2}$ state is found in the trigonally distorted $[Fe(H_2O)_6]SiF_6$ for which $e^2qQ$ is negative. One of the few known examples of a doublet ground state is in trigonally distorted $FeCO_3$, and the low-temperature quadrupole splitting value of +2.06 mm $s^{-1}$ is about half that of $-3.61$ mm $s^{-1}$ for the fluorosilicate. All three compounds have a distorted 6-coordination to oxygen. The quadrupole splitting originates *not* from asymmetric covalency with the ligands, but from the sensitivity of the orbital state of an essentially non-bonding electron to the geometrical environment. This effect provides a powerful means of studying small distortions from regular co-ordination.

At an infinitely high temperature, all the levels of the $t_{2g}$ multiplet would be equally populated by thermal excitation, and the electric field gradient would once again be zero. At intermediate temperatures, a partial degree of thermal excitation will cause a partial cancellation of the electric field gradient, so that the quadrupole splitting should decrease with rise in temperature. A full mathematical treatment of the temperature dependence, (an outline of which is given below) was first given by Ingalls [1].

The effects of the ligand field on the free-ion wavefunctions are calculated using a perturbation Hamiltonian

$$\mathscr{H} = v_0 + v_T + v_R + \lambda\hat{L}.\hat{S}$$

where the successively smaller terms are $v_0$ the axial field, $v_T$ the tetragonal (or trigonal) distortion, $v_R$ the rhombic distortion, and $\lambda\hat{L}.\hat{S}$ the spin-orbit coupling. This last term has a considerable effect on the quadrupole splitting at low temperatures because it causes admixture of excited states into the ground state. The free-ion $^5D$ state of $Fe^{2+}$ has $S = 2$ and $L = 2$ so that there are $(2S + 1)(2L + 1) = 25$ energy states in the ligand field. Some typical energy level schemes from such calculations are illustrated in Fig.5.1 for the lowest 15 levels only.

If the energy separation of a given excited level is $E_j$, its relative population is given by Boltzmann's theory as $e^{-E_j/kT}$ for a ground state population of unity. The thermal averaging takes place within a time-scale less than the Mössbauer event (ca. 100 ns for $^{57}Fe$), so that the total resultant quadrupole splitting will be a statistically averaged quantity. Each excited level has an electric field gradient represented by $V^j_{zz} = eq_j$ and $\eta_j$, and the quadrupole splitting at any temperature $T$ is given by

$$\Delta_T = \tfrac{1}{2}e^2Q \frac{\left[\left(\sum_j q_j e^{-E_j/kT}\right)^2 + \tfrac{1}{3}\left(\sum_j \eta_j q_j e^{-E_j/kT}\right)^2\right]^{1/2}}{\sum_j e^{-E_j/kT}}$$

This summation is made over all 25 levels. In the event that $v_T$ is much greater than $v_R$ and $\lambda\hat{L}.\hat{S}$, the latter terms may be ignored. The problem then reduces to one of thermal excitation from a pure $d_{xy}$ ground state to the $d_{xz,yz}$ excited state at an energy $E_0$, and results in a very simple expression for the temperature dependence of

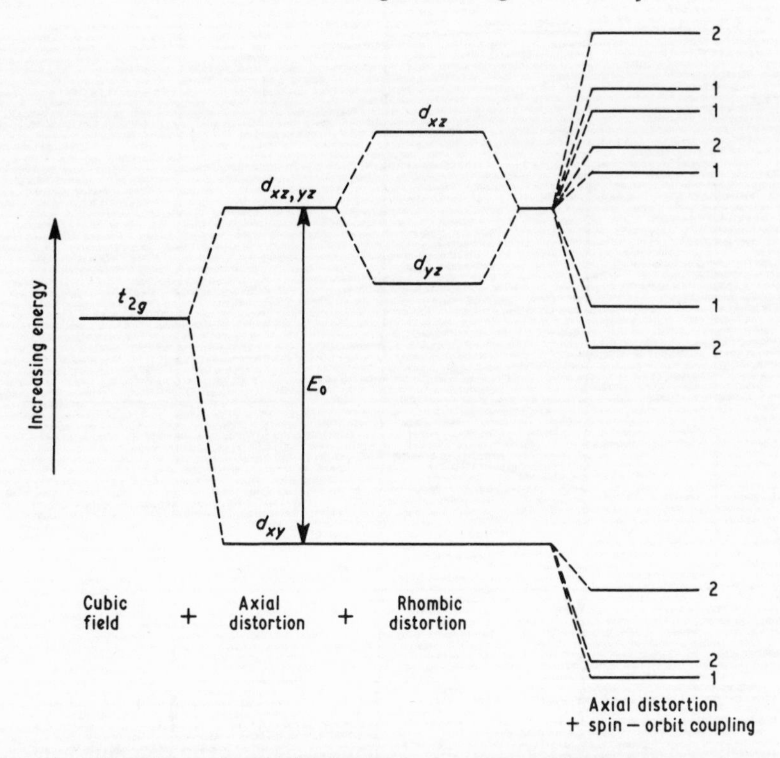


Fig. 5.1 The effect of axial (tetragonal), rhombic, and spin-orbit coupling perturbations on the $t_{2g}$ energy levels of the $Fe^{2+}$ ion. The level scheme on the right is a combined axial distortion + spin-orbit coupling, and the numbers indicate the degeneracy of the states.

$$\Delta_T = \Delta_0 \frac{(1 - e^{-E_0/kT})}{(1 + 2e^{-E_0/kT})}$$

where $\Delta_0$ is the quadrupole splitting due to a single $3d$-electron in the $d_{xy}$ state (~4 mm $s^{-1}$).

This function is approximately valid except at low temperatures where the spin-orbit coupling causes a reduction from the value $\Delta_0$. The effects of the axial field and spin-orbit coupling when the full calculation is made [2] are illustrated in Fig.5.2. The curves represent the observed quadrupole splitting $\Delta_T/\Delta_0 = F$ for a $t_{2g} - e_g$ separation of 10 Dq = 10 000 $cm^{-1}$ (=1.9878 x $10^{-19}$ J ≡ 119.7 kJ $mol^{-1}$) and a spin-orbit coupling of $\lambda = -80$ $cm^{-1}$. The numerical figures refer to the total splitting of the $t_{2g}$ levels,

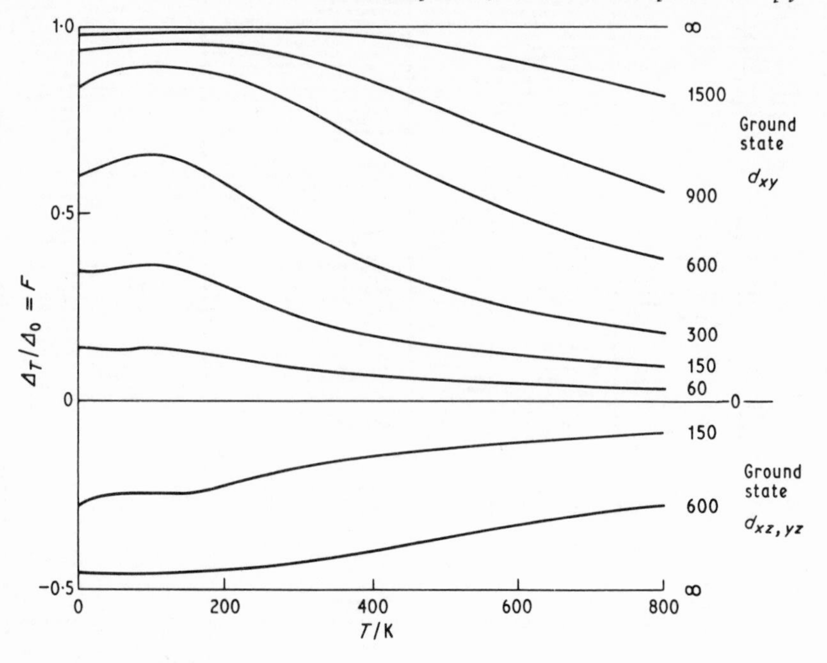


Fig. 5.2 Calculated values for the quadrupole splitting $\Delta_T/\Delta_0 = F$ in units of the value for a 3*d*-electron. The curves are drawn for different values of the splitting of the $t_{2g}$ levels (in $cm^{-1}$), and for a spin-orbit coupling parameter of $\lambda = -80\ cm^{-1}$.

$E_0$ in $cm^{-1}$, by the axial field. A positive value of $F$ corresponds to a singlet ground state and a positive value for $e^2qQ$, while a negative value corresponds to a doublet ground state and a negative value for $e^2qQ$ of approximately half the magnitude. Note in particular the large reduction in $\Delta_T$ at zero temperature which is a direct result of the spin-orbit coupling.

The magnitude of the quadrupole splitting is very sensitive to the axial (or trigonal) distortion and the spin-orbit coupling, but not to the rhombic distortion which has a greater effect on the less-populated excited states. The presence of a rhombic distortion does however cause the asymmetry parameter to be non-zero and temperature dependent, and an example of this can be found in $[Fe(H_2O)_6](NH_4SO_4)_2$ where the large rhombic field causes state-mixing with the nominal $d_{xy}$ ground state so that $\eta = 0.3$ at 4.2 K and 0.7 at 300 K [3].

Although the above theory is precise, there are a number of additional factors which must also be considered. Firstly, the value of $\Delta_0$ is dependent on the value of $\langle r^{-3} \rangle$ for a 3*d*-electron, and this effectively decreases in the presence of covalent bonding. The upper limit for $\Delta_0$ is about 4.0 mm $s^{-1}$ in ionic compounds, but is not known accurately. A second factor which is not so easy to estimate is the contribution to the electric field gradient tensor from charges external to the central atom, the so-called 'lattice' term. Ingalls considers the lattice correction to be of opposite sign to that of the 3*d*-orbital contribution [1], but direct lattice-sum calculations from known crystal structures have suggested that there need be no direct relationship between the two [4]. The lattice-term is neglected in many instances on the grounds that it appears to be usually less than 10% of the total field gradient, but there have been recent reports of some $Fe^{2+}$ compounds with unusually large quadrupole splittings {e.g. 4.65 mm $s^{-1}$ at 300 K in $Fe(NH_2CSNHNH_2)_2SO_4$ which is also unusual in having two geometrically distinct iron sites, both with six co-ordination, and with identical Mössbauer parameters [5]} and this has raised fresh conjecture on the problem.

A third major aspect is the possibility of temperature dependence of the ligand field splittings and the lattice-term as a direct result of thermal expansion. A major attempt has been made to include all these corrections in the analysis of data from $FeF_2$ between 78 K and 1023 K [6]. The experimental data and the final computed fit are shown in Fig.5.3. However it is generally acknowledged that such sophisticated theoretical models are unlikely to provide a unique interpretation of the experimental data.

Of the other common paramagnetic oxidation states of iron, tetrahedral high-spin $Fe^{2+}$ and octahedral low-spin $Fe^{III}$ can be treated in an analogous manner. The *E* ground state of $Fe^{2+}$ in a cubic tetrahedral field is also split by an axial distortion, but the spin-orbit coupling is only effective in second-order terms and can be neglected. The resulting thermal population of the $d_{z^2}$ and $d_{x^2-y^2}$ levels results in a limiting value of the quadrupole splitting at zero-temperature which is not reduced by the spin-orbit coupling and is independent of the axial field. Thus the observed variation [7] in the values at 4.2 K for $FeCl_4^{2-}$ (3.27 mm $s^{-1}$), $FeBr_4^{2-}$ (3.23), $Fe(NCS)_4^{2-}$ (2.83), and $Fe(NCSe)_4^{2-}$ (2.73) can be accounted for solely in terms of a difference in

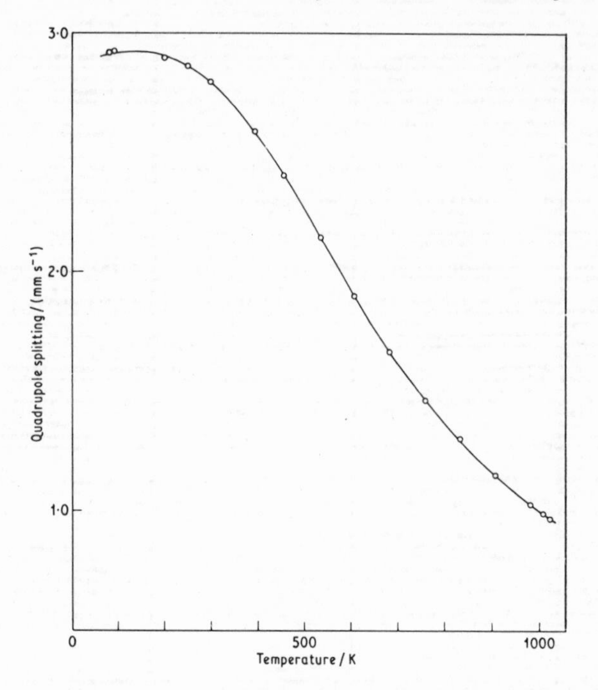


Fig. 5.3 The temperature dependence of the quadrupole splitting in $FeF_2$ with a theoretical curve from a model incorporating thermal lattice expansion. ([6], Fig. 2).

lattice term or a change in $\langle r^{-3}\rangle$ due to covalency. The temperature dependence relationship is particularly simple in this instance, being

$$\Delta_T = \Delta_0 \frac{(1 - e^{-E_0/kT})}{(1 + e^{-E_0/kT})}$$

$$= \Delta_0 \tanh(E_0/2kT)$$

where $E_0$ is the separation of $d_{z^2}$ from $d_{x^2-y^2}$.

Octahedral $Fe^{III}$ has a $^2T$ ground state ($t_{2g}^5$) and corresponds to a single electron-hole in an otherwise cubic triplet level. It is mathematically similar to octahedral $Fe^{2+}$ except that because $S = \frac{1}{2}$ there are only six sub-levels to consider in the $t_{2g}$ multiplet, and the spin-orbit coupling parameter is much larger. The available data confirm the expected temperature dependence, but the effect of the greatly increased covalency on the value of $\langle r^{-3} \rangle$ has not been established satisfactorily. Similar arguments can be used for other elements. However in the second and third-row transition metals the spin-orbit coupling constant is much larger, and may result in an almost temperature-independent and much reduced quadrupole splitting below 300 K.

Not all paramagnetic configurations of iron show a temperature dependent quadrupole splitting. The high-spin $Fe^{3+}$ ion is in a $^6S$ state ($t_{2g}^3 e_g^2$) which has spherical symmetry and therefore no inherent electric field gradient. Any observed splitting arises solely from charges external to the cation, either by direct contribution or by indirect polarization. Point charge summations of the lattice contribution correlate approximately with the observed splitting, but tend to be unrealistic because the major contributions arise very close to the central atom where the point charge model is particularly inadequate.

The principles outlined above can also be applied to the less common configurations of iron, but with varying results. For example the quadrupole splitting in phthalocyanine iron(II) is 2.70 mm $s^{-1}$ at 4.2 K and only decreases to 2.62 mm $s^{-1}$ at 293 K. The iron has two unpaired electrons ($S = 1$), being in an intermediate spin-state appropriate to the square-planar geometry. The lack of orbital degeneracy and low-lying excited states in this configuration prevents significant temperature dependence. In consequence, very little information is obtained from the quadrupole splitting *in isolation*. However, in such cases it is often possible to make more positive deductions from a combination of quadrupole and magnetic hyperfine data.

## 5.2 Magnetic hyperfine interactions

It was shown in Chapter 2 that the presence of a magnetic hyperfine field causes line splitting, and it was intimated that this field could be intrinsic to the resonant atom. If the atom has one or more unpaired *d*- or *f*-electrons, it is intuitively obvious that their interaction with the *s*-electrons of parallel spin will be different to that with the *s*-electrons of opposed spin. This interaction results in a slight imbalance of spin density at the nucleus [8], thereby generating a magnetic field with a flux density which can be expressed as

$$B_S = \left(\frac{2\mu_0}{3}\right) \mu_B \left\{ |\psi_\uparrow(0)|^2 - |\psi_\downarrow(0)|^2 \right\}$$

where $\mu_0$ is the permeability of a vacuum ($4\pi \times 10^{-7}$ kg m s$^{-2}$ A$^{-2}$) and $\mu_B$ is the Bohr magneton ($9.273 \times 10^{-24}$ A m$^2$). $|\psi_\uparrow(0)|^2$ represents the electron density at the nucleus with spin parallel to the magnetic moment, and $|\psi_\downarrow(0)|^2$ the antiparallel spin-density. B therefore has the units of tesla (T) or kg s$^{-2}$ A$^{-1}$ (1 T = 1 Wb m$^{-2}$ = $10^4$ G where Wb denotes Weber and G denotes Gauss). $B_S$ is usually referred to as the Fermi contact term, and is directly proportional to the total spin-imbalance of the *s*-electrons with a magnitude which can be as high as 200 T. It is not, however, the only possible intrinsic contribution to the field.

If there is a non-zero orbital magnetic moment on the resonant atom, then there is a further contribution to the flux density

$$B_L = \left(\frac{\mu_0}{2\pi}\right) \mu_B \langle r^{-3} \rangle \langle L_z \rangle$$

where $\langle r^{-3} \rangle$ is the expectation value in m$^{-3}$ of $1/r^3$ for the electrons producing the moment $\langle L_z \rangle$. The orbital moment $\langle L_z \rangle$ is normally quenched in for example $Fe^{2+}$ ions, but spin-orbit coupling can induce a flux density of

$$B_L = \left(\frac{\mu_0}{2\pi}\right) \mu_B \langle r^{-3} \rangle (g_z - 2) \langle S_z \rangle$$

where $\langle S_z \rangle$ is the expectation value of the spin moment, and $g_z$ is the electronic Landé splitting factor. In high-spin $Fe^{2+}$ compounds where $S = 2$ the orbital contribution, $B_L$, is opposite in sign to $B_S$ and varies from about +20 to +60 T.

A third major contribution to the magnetic flux density arises from a dipolar interaction of the spin moment of the atom with the nucleus, which for axial symmetry is

$$B_{\mathrm{D}} = -\left(\frac{\mu_0}{8\pi}\right) \mu_{\mathrm{B}} \langle r^{-3} \rangle \langle 3 \cos^2 \theta - 1 \rangle \langle S_z \rangle$$

This expression is closely related to that for the quadrupole splitting $e^2qQ$. For example, a high-spin $Fe^{2+}$ ion has a total spin of $S = 2$; it has an electric field gradient $V_{zz} = eq$ given by writing equation 2.19 in alternative form as

$$4\pi\epsilon_0 q = -\langle r^{-3} \rangle \langle 3 \cos^2 \theta - 1 \rangle$$

so that

$$B_{\mathrm{D}} = \left(\frac{\mu_0 \epsilon_0}{2}\right) \mu_{\mathrm{B}} q \langle S_z \rangle$$

Estimated values for $B_{\mathrm{D}}$ are usually in the range 0–8 T, and $B_{\mathrm{D}}$ is of course zero in cubic symmetry.

The total magnetic flux density can therefore be written as

$$B = B_0 + B_{\mathrm{S}} + B_{\mathrm{L}} + B_{\mathrm{D}}$$

where $B_0$ is the flux density arising from an externally applied field (not necessarily collinear with $B_{\mathrm{S}}$ etc.). From the foregoing discussion one might expect that all compounds with unpaired valence electrons would show a magnetic splitting, but this is not so. The Hamiltonian in equation 2.9 contains **I.B** as a vector product, and the Mössbauer event in for example $^{57}$Fe lasts for about 100 ns. In the absence of an externally applied field, the electronic spin which generates **B** will normally be undergoing electronic spin relaxation, and if the relaxation rate is far faster than 100 ns, the time-average $\langle \mathbf{B} \rangle$ is zero. This situation applies for example to paramagnetic $Fe^{2+}$ ($S = 2$) ions, so that the Mössbauer spectrum shows only a quadrupole interaction. The relaxation rate is much slower in $Fe^{3+}$ ($S = \frac{5}{2}$) ions and in some of the rare-earth ions, and in these cases one can see magnetic splitting. However, as these also frequently embrace the intermediate case where both the relaxation and Mössbauer processes operate on the same timescale, discussion of these is deferred until Chapter 6 on dynamic effects.

A further possibility is that the spins on neighbouring ions are strongly interacting so as to have a preferred direction, and that this leads to ferromagnetic or antiferromagnetic coupling. The nucleus sees a time average $\langle S_z \rangle$ of the electronic spin $S$ which is non-zero. The relaxation rates for $\langle S_z \rangle$ in such cooperative phenomena are generally very much slower, and magnetic hyperfine splitting is seen without dynamic effects. The internal magnetic field at any given site in a magnetically ordered material is generally proportional to the magnetization at that site. The temperature dependence of the magnetization usually follows a Brillouin function, which may be written as

$$\frac{M_T}{M_0} = B_S \left( \frac{3S}{S+1} \cdot \frac{T_C}{T} \cdot \frac{M_T}{M_0} \right)$$

where

$$B_S(x) = \frac{2S+1}{2S} \coth \left( \frac{2S+1}{2S} x \right) - \frac{1}{2S} \coth \left( \frac{1}{2S} x \right)$$

$M_T$ is the magnetization at a temperature $T$, $M_0$ is the saturation magnetization, and $T_C$ is the Curie temperature. The function is illustrated in Fig.5.4 for a value of $S = 1$ in terms of the reduced variables $M_T/M_0$ and $T/T_C$. The magnetization and hence the internal field decrease with temperature rise and become zero at the Curie or Néel temperature. Whereas the bulk magnetization is an average effect of all magnetic sites in the crystal, the Mössbauer spectrum is capable of distinguishing the magnetization at more than one type of crystal site. This is particularly useful in the study of antiferromagnetic compounds where the bulk magnetization is unduly sensitive to ferromagnetic impurities.

Although several Mössbauer isotopes have been used to study magnetic ordering in intermetallic alloy systems (see Chapter 8) the only nuclide which commonly shows magnetic ordering in coordination compounds is $^{57}Fe$. The simplest case to consider is the $Fe^{3+}$ ($S = \frac{5}{2}$) ion, in for example $FeF_3$ which is antiferromagnetic below 363 K. The $S$-state ion is in a nearly octahedral environment, and above 363 K shows a single resonance because $\langle S_z \rangle$ time-averages to zero [9]. Below this temperature, the co-operative ordering makes $\langle S_z \rangle$ non-zero so that the flux density due to the

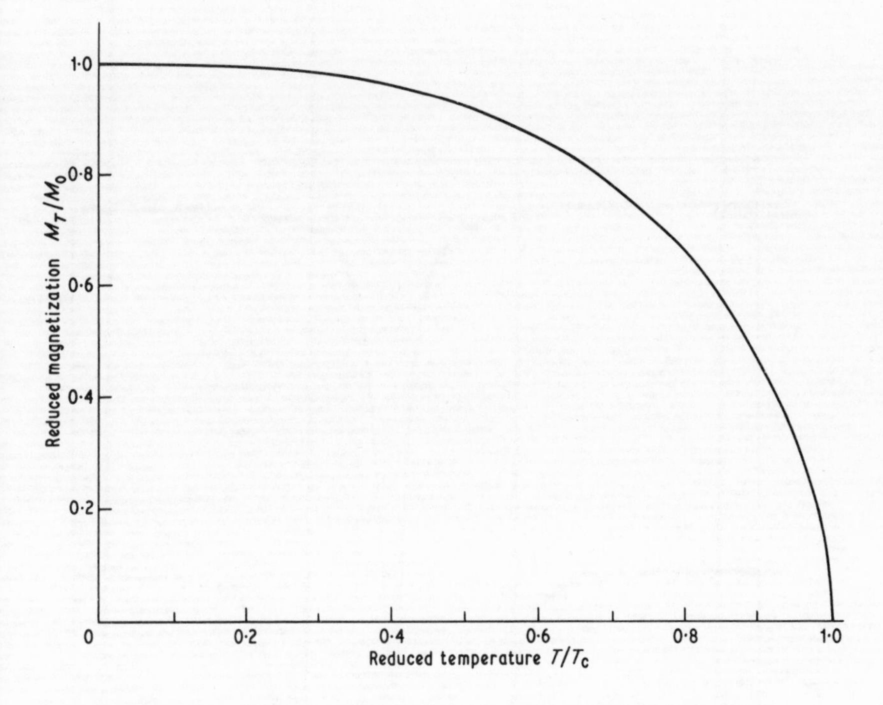

Fig. 5.4 The $S = 1$ Brillouin function in terms of the reduced variables $M_T/M_0$ and $T/T_C$.

contact field, $B_S$, becomes non-zero, and a six-line magnetic hyperfine splitting appears (Fig.5.5). The magnetic flux density increases very rapidly at first, approximately following the $S = \frac{5}{2}$ Brillouin function, with a saturation value at zero temperature of 62 T (620 kG). The saturation value is found to vary considerably from compound to compound, being 55 T in $Fe_2(SO_4)_3$, 47 T in $FeCl_3$ and $FeCl_4^-$ and 42 T in $FeBr_4^-$. A decrease in flux density corresponds to an increase in covalency, and represents an effective decrease in polarization of the Fe inner core as the spin-density is delocalized towards the ligands. As $B_L$ and $B_D$ are always small for the $S$-state $Fe^{3+}$ ion, they can be neglected to first order. A further noteworthy feature of the $Fe^{3+}$ ion is that it has a very low magnetic anisotropy. Application of an external magnetic field causes the intrinsic magnetization to rotate into the axis of the external field, and the sum $B = B_0 + B_S$ becomes algebraic in the manner described in Chapter 2. This feature will

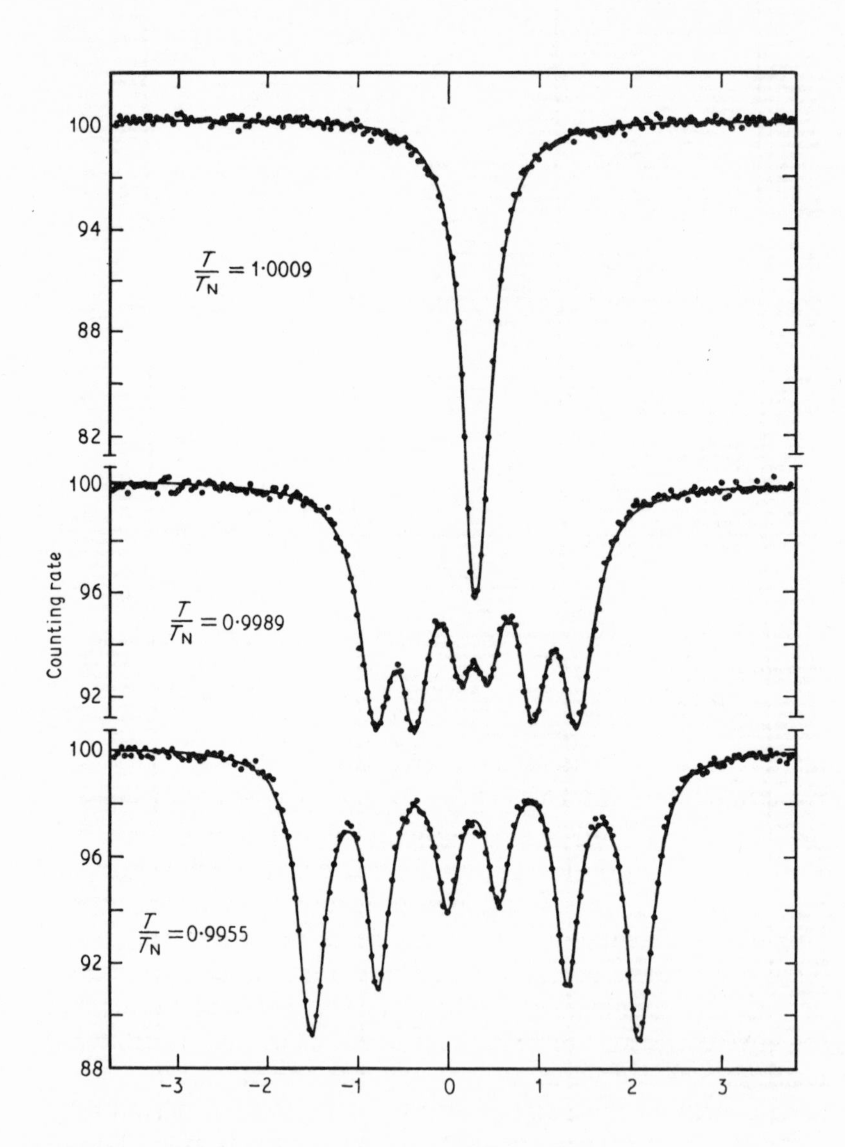


Fig. 5.5(a) The Mössbauer spectrum of $FeF_3$ in the vicinity of the Néel temperature. ([9], Fig.2)

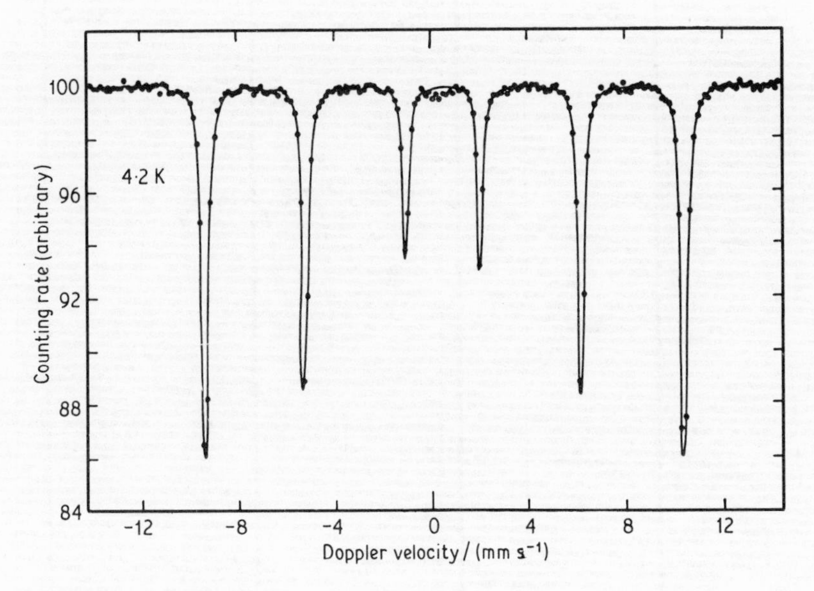


Fig. 5.5(b) The Mössbauer spectrum of $FeF_3$ at 4.2 K. ([9], Fig.1)

be mentioned again more specifically in connection with iron oxides in Chapter 7.

Excluding the variations due to covalency, the value of $B_S$ for any configuration of iron is given approximately (in tesla) by $-22\langle S_z \rangle$. Measurements in an applied external field show the sign to be negative in all cases. The flux density in $Fe^{3+}$ ($S = \frac{5}{2}$) compounds averages about 55 T, while $B_S$ for an $Fe^{2+}$ ($S = 2$) compound is only 44 T.

The total saturation flux density in ordered $Fe^{2+}$ compounds is very variable, being $-32.9$ T in $FeF_2$, $-25$ T in $FeCl_2.2H_2O$ and $+3$ T in $FeBr_2$. This is because the values of $B_L$ and $B_D$ may be quite large for this electron configuration. Although $B_S$ and $B_L$ are usually much greater than $B_D$, they are usually in opposition to each other so that the resultant flux density can be quite small and of either sign.

An example of magnetic hyperfine splitting from a high-spin $Fe^{2+}$ ion is given by data for $FeF_2$ [10]. Typical spectra are illustrated in Figs. 5.6 and 5.7. At 78.21 K which is 0.09 degrees

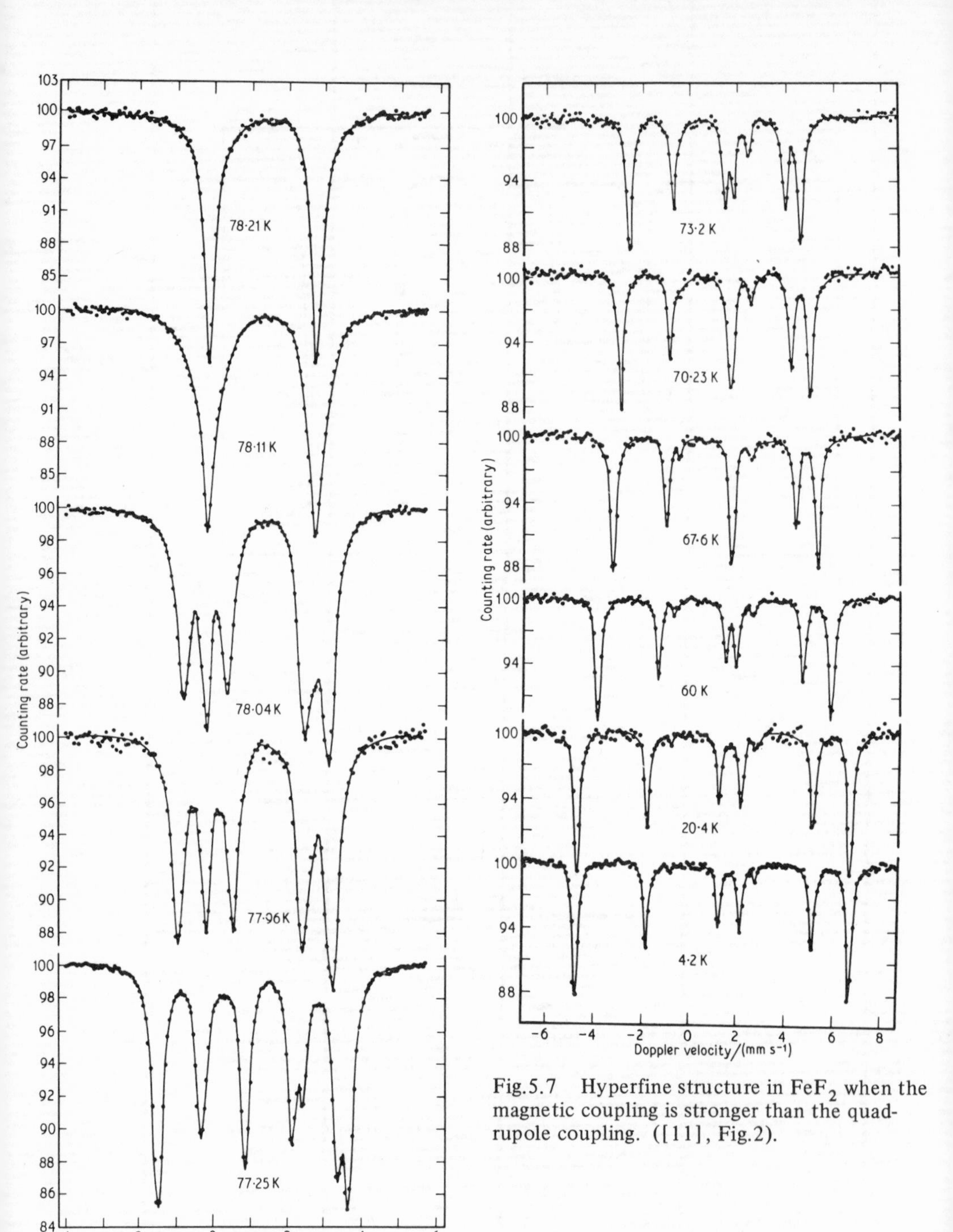


Fig.5.7 Hyperfine structure in $FeF_2$ when the magnetic coupling is stronger than the quadrupole coupling. ([11], Fig.2).

Fig. 5.6 Magnetic hyperfine structure of $^{57}Fe$ in $FeF_2$ within 1 K of the Néel temperature. The quadrupole coupling is stronger than the magnetic coupling. ([10], Fig. 1).

above the Néel temperature a sharp doublet is seen. At 78.11 K, only 0.01 degrees below the critical point, considerable line broadening is found, and the subsequent splitting resembles the triplet-doublet pattern found for magnetic perturbation of a diamagnetic compound by an applied field. This is because the principal axis of the electric field gradient in $FeF_2$ is perpendicular to the spin axis and thus approximates to the most probable relationship between an external field and the electric field gradient in a powdered sample. At low temperatures all six lines of the spectrum are clearly resolved, and there are indications of weak absorption in the two formally forbidden lines ($\Delta m_z = \pm 2$) because in this example of combined magnetic/quadrupole interactions the energy levels are no longer pure eigenstates of $m_z$. The line positions and relative intensities contain far more information than is available from the quadrupole spectrum alone. Mössbauer spectra from single crystals show that the magnetic spin axis is the crystal '*c*' axis. The data illustrated at 4.2 K lead to a value for $B$ of 32.9 T, a value for $\Delta = \frac{1}{2}e^2qQ$ of +2.85 mm $s^{-1}$, and an asymmetry parameter of $\eta = 0.40$. The line positions and intensities require that the principal axis of the electric field gradient tensor is perpendicular to the spin axis which is itself along the minor axis. Thus analysis of a combined magnetic/quadrupole interaction gives detailed information about the crystal symmetry and the magnetic structure.

It is still possible to use an applied magnetic field perturbation to measure the sign of $e^2qQ$ in a paramagnetic $Fe^{2+}$ compound, but under more restrictive conditions. At moderate temperatures the spin relaxation remains fast and the method works. At very low temperatures the magnetic susceptibility becomes high, the electrons tend to align with the applied field, and the observed magnetic splitting may not only be far greater than that due to the applied field alone, but becomes more difficult to interpret [11]. The anisotropy of the $Fe^{2+}$ ion environment must be considered in detail.

The other electronic configurations of iron all have a lower value of $S$, and although for example $K_3[Fe(CN)_6]$ with $S = \frac{1}{2}$ shows antiferromagnetic ordering, the Néel temperature is only 0.13 K. Co-operative phenomena as such are therefore not easy to study. In paramagnetic configurations where the geometry is highly distorted such as phthalocyanine iron(II) (square-planar,

$S = 1$) and bis-(N,N-diethyldithio-carbamato)iron nitrosyl (approximately square-based pyramidal, $S = \frac{1}{2}$), the iron environment is highly anisotropic. The quadrupole splitting observed is usually temperature independent because the ligand field splitting of the non-bonding $3d$-orbitals is much larger than in the more regular complexes. It therefore gives little information when taken in isolation. Application of a magnetic field at very low temperatures can induce a large magnetic splitting in a similar manner to $Fe^{2+}$, and makes it possible to determine the electronic ground state.

A good example of this is given by the $S = \frac{1}{2}$ nitrosyl complex just mentioned [12]. The molecular unit $(Et_2NCS_2)_2Fe(NO)$ is formally an iron(I) complex if the nitrosyl is considered to be $\dot{N}O^+$. In attempting to deduce the electronic configuration the chemical isomer shift is uninformative because there are no comparative data for iron(I) systems. The quadrupole splitting decreases from 1.03 mm s$^{-1}$ at 1.3 K to 0.89 mm s$^{-1}$ at 300 K. The sign of $e^2qQ$ is positive, but it requires a detailed analysis of measurements in applied magnetic fields to establish unequivocally that it is an $S = \frac{1}{2}$ complex with a $d_{z^2}$ ground state. As such it would be expected to have a negative sign for $e^2qQ$ were it not for the large contributions to the electric field gradient from other delocalized bonding electrons and from lattice contributions, which in the event cause a change of sign.

Diamagnetic ions do not show intrinsic magnetic splitting, but in some circumstances the presence of a nearby magnetically-aligned ion can induce a small spin polarization resulting in an effect known as transferred hyperfine splitting. This is known in for example tin, antimony, tellurium and iodine, but will be considered more explicitly in Chapter 7.

## 5.3 Spin cross-over

In the majority of paramagnetic transition metal compounds it is sufficient to distinguish between a small ligand-field splitting of the $d$-orbitals and thence a 'high-spin' configuration, and a large ligand-field splitting which induces spin-pairing to give a 'low-spin' configuration. However, in a small minority of compounds one finds that both configurations can exist, sometimes even in apparently stable coexistence, and these are known as spin-crossover complexes. In octahedral geometry the phenomenon is limited to $d^4$, $d^5$, $d^6$ and $d^7$ configurations, and is far from understood. For this reason Mössbauer spectroscopy has been extensively used to study examples of spin-crossover in iron compounds.

In iron(II) complexes the change is from a $^5T_2$ ($S = 2$) state to the $^1A_1$ ($S = 0$) state. These configurations are particularly distinct in the Mössbauer spectrum and their relative proportions can be determined easily. Most iron(II) complexes with 1,10-phenanthroline are in either the $^5T_2$ or $^1A_1$ configuration, but in two instances a cross-over has been found, $Fe(phen)_2(NCS)_2$ and $Fe(phen)_2(NCSe)_2$. At high temperatures the magnetic moments are about 5.2 $\mu_B$, typical of the high-spin state, but at 174 K and 232 K respectively there is a comparatively sudden change to the low-spin state, although the low-temperature form usually has a small residual moment which varies from sample to sample [13]. If the cross-over were to involve a simple thermal population of close-lying $^5T_2$ and $^1A_1$ states, there would be no sudden change. In other compounds such as $Fe(bipy)_2(NCS)_2$ and complexes with 2-aminomethylpyridine, 2-(2′-pyridyl)imidazole and hydro-tris-(1-pyrazolyl)borate ligands, the transition is equally abrupt. In the region of change-over both spin-states appear to co-exist, and this is particularly marked in the borate complex as illustrated in Fig.5.8, where both states coexist over a range of 50 degrees but show most change over only a few degrees [14]. In this instance at least, the effect seems to show signs of hysteresis as well as a dependence on particle size, and slow thermal cycling of a crystal causes pulverization.

In several cases there is good evidence for substantial structural change involving alterations in molecular dimensions. The equilibrium bond distances in the $^5T_2$ and $^1A_1$ forms are unlikely to be identical because of different electron correlation, and this could induce crystallographic changes. Furthermore, some of the large ligands involved could, in principle, form different geometric isomers. Such changes would be consistent with the apparent slowness of the interconversion ($> 10^{-7}$ s) as deduced from the Mössbauer spectrum, and with the retention of a small proportion of one form well below the nominal transition temperature.

The iron(III) cross-over from $^6A_1$ ($S = \frac{5}{2}$) to $^2T_2$ ($S = \frac{1}{2}$) is less common and is best known in the N,N-dialkyldithiocarbamates such as $(Me_2NCS_2)_3Fe$. These compounds do not usually show

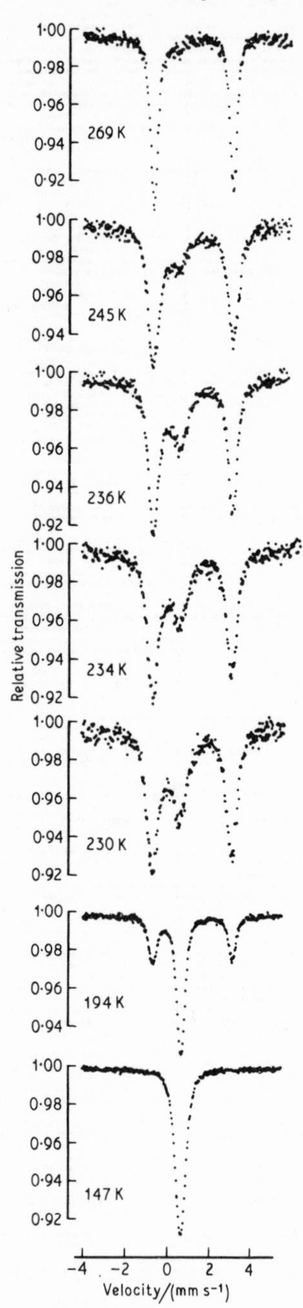


Fig. 5.8 Typical spectra of a hydro-tris-(l-pyrazolyl)borate iron(II) complex showing the $^5T_2$-$^1A_1$ crossover. ([14], Fig. 5).

any sudden discontinuity in the magnetic susceptibility curve, and in the case of the methyl compound the susceptibility rises steadily from about 1.8 $\mu_B$ at 4.2 K to about 5 $\mu_B$ at 400 K. At temperatures above 78 K the Mössbauer spectra comprise a simple, poorly resolved doublet [15]. There are no obvious signs of two components, and although the lines broaden on cooling, this can if desired be ascribed to spin relaxation (see Chapter 6) which in any case becomes clearly significant below 78 K. The evidence seemed at first to favour a fast rate of exchange between the $^6A_1$ and $^2T_2$ states. However, recent data for some tris(monothio-β-diketonato)iron(III) complexes [16] have shown clear evidence for the coexistence of both high-spin and low-spin species during the Mössbauer lifetime. Whether by chance the two species are unresolved in the former case has not been ascertained. Another problem raised by the new data concerns the relative proportions of the two species. If the high-spin form is achieved by thermal excitation from a $^2T_2$ to a $^6A_1$ state, then even at very high temperatures the proportion of the $^6A_1$ form can never exceed 50%. However, in the diketonato complexes there is evidence to suggest that a more complete transfer to the $^6A_1$ state takes place.

Mössbauer spectroscopy is particularly useful in these studies of spin-crossover, because, unlike magnetic susceptibility measurements, it is not an averaged measurement but records uniquely the different species present.

## 5.4 Pressure effects

The convenience of carrying out experiments at atmospheric pressure has led to a tendency to measure only the temperature dependence of physical properties. However, it is equally feasible in principle, if technically difficult in practice, to observe phenomena which are pressure-dependent. This applies to both diamagnetic and paramagnetic materials, but as the most interesting effects concern the latter, they are conveniently discussed here.

Extreme pressure (up to 200 000 atmospheres; 1 atm = 1.01325 $\times 10^5$ Pa = 1.01325 bar) will compress a material, which in the context of $^{57}$Fe can be thought of as an increase in the overlap between the inner *s*-shells of iron and the valence orbitals of the ligands. This generates an overlap-induced increase in *s*-electron density at the nucleus, and a decrease in the chemical isomer shift.

This decrease in shift with increasing pressure has been observed in a wide range of compounds by Drickamer and co-workers [17], but has only been fully confirmed with wave-function calculations in the case of $KFeF_3$ [18]. The effect is of course fully reversible on release of the pressure.

If the iron environment is non-cubic so that a quadrupole splitting is present, the latter also changes with pressure, but the behaviour is less easy to rationalize because of the possibility of pressure-induced distortions.

The most interesting aspect of pressure studies is the tendency for iron in many compounds to show a change in oxidation state or electronic configuration at high pressures. This usually affects only a fraction of the total iron, and is fully reversible on pressure release. The most common instance of this is found in high-spin iron(III) compounds such as $Fe_2(SO_4)_3$ and $FePO_4$ which show a partial conversion to high-spin iron(II). Low-spin iron(III) can change to low-spin iron(II) as in $K_3Fe(CN)_6$. Such a change in oxidation state results from an electron transfer from ligand to metal. Compression alters the spacing of the electronic energy levels of the system in such a way that thermal excitation can cause electron transfer to a level which was inaccessible under ambient conditions. $K_3Fe(CN)_6$ proves particularly interesting because there is also a first-order phase transition at about 50 kbar which presumably relieves the internal compression of the hexacyanoferrate(III) ion and causes a large decrease in the observed fraction of the reduced species.

A particularly clear example of high-spin to low-spin conversion is given in Fig.5.9 which shows the effect of pressure on the Mössbauer spectrum of $^{57}Fe$ impurity atoms in $MnS_2$ (the impurity concentration was 2 at.%.).$MnS_2$ is isostructural with $FeS_2$ in which the iron is in the low-spin configuration. The impurity atoms in $MnS_2$ are in the high-spin configuration at atmospheric pressure because the lattice constant of the pyrite lattice has increased to 6.102 Å in $MnS_2$ compared to 5.504 Å in $FeS_2$. At comparatively modest pressures the impurity atoms revert to the low-spin configuration found in the iron analogue [19].

The occurrence of such changes is often intimately linked to the $\pi$ and $\pi^*$ orbitals on the ligands and the changing ability of the metal to back-donate to the $\pi^*$ system as the overlap and energy separations alter. That the balance is a subtle one can be

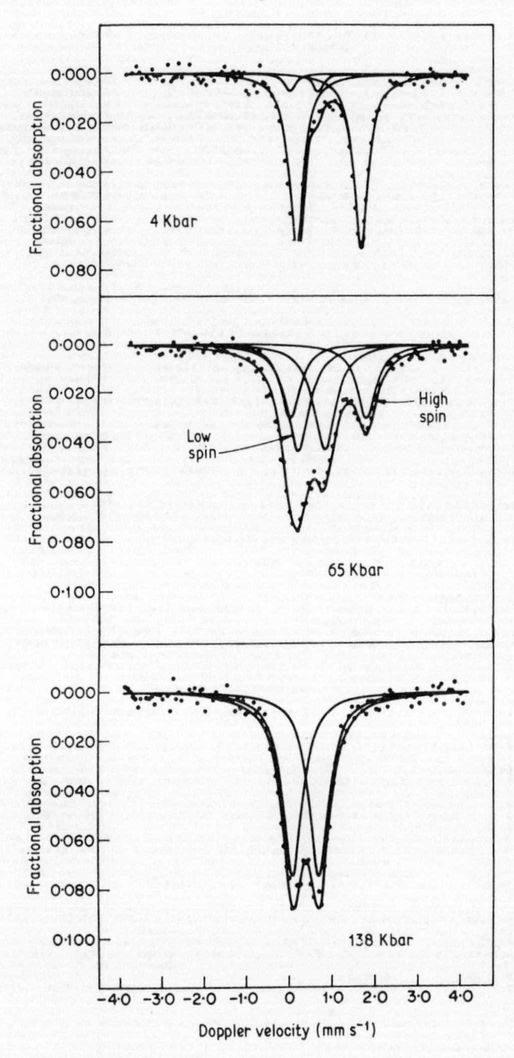


Fig. 5.9 The $^{57}$Fe Mössbauer spectra of $MnS_2$ containing 2 at.% $^{57}$Fe impurity atoms under a pressure of 4, 65 and 138 kbar. Note the change from high-spin $Fe^{2+}$ to low-spin $Fe^{2+}$ with increase in pressure. ([19], Fig.1).

seen from the fact that high-spin iron(II) in bis-phenanthroline complexes converts to low-spin iron(II) with increase in pressure [20], whereas low-spin iron(II) in tris-phenanthroline complexes converts to high-spin iron(II).

## 5.5 Second and third row transition elements

The compounds of the second and third row transition elements differ significantly from those of the first row. The ligand-field effects on the *d*-electrons are usually much stronger, and result in more frequent adoption of low-spin configurations. The magnetic properties are also more complicated. In the second row, significant Mössbauer data are only available for ruthenium, which is of particular interest because of its relationship to iron in the Periodic Table. Many of the third row elements have at least one known Mössbauer resonance, and although none have been studied extensively as yet, the main features of several of them are now well established.

The 90-keV resonance of $^{99}$Ru is comparatively difficult to observe, partly because the high $\gamma$-ray energy makes the resonance weak except at liquid helium temperature, and partly because of the short 16-day half-life of the $^{99}$Rh parent. However, with the recent improvements in experimental technique, several laboratories have obtained good data for a wide range of compounds. Although the transition is from the $I_e = \frac{5}{2}$ ground state to the $I_e = \frac{3}{2}$ excited state, the quadrupole moment of the ground state is very small. As a result, any quadrupole effect is usually seen as an apparent doublet splitting of the excited state.

The quadrupole splitting and chemical isomer shift data at 4.2 K for a small selection of compounds are given in Table 5.1 [21, 22]. As with iron, the ruthenium oxidation states are characterized by the *d*-electron configuration. The chemical isomer shift decreases as the number of 4*d*-electrons increases, which by analogy with iron indicates $\delta R/R$ to be positive. Within the range of shifts shown by Ru(II) compounds, the higher values for the cyanide and nitrosyl derivatives results from back-donation to the ligand $\pi^*$ orbitals. This behaviour is comparable with that found in the iron cyanides. In general the chemical isomer shifts correlate with the spectrochemical series of the ligands.

Diamagnetic compounds with a regular geometry such as $RuO_4$, $[Ru(NH_3)_6]^{2-}$ and $[Ru(CN)_6]^{4-}$ show no quadrupole splitting. However, $RuO_4^-$ and $RuO_4^{2-}$ with one and two unpaired electrons respectively have a distorted geometry in analogy with $[FeX_4]^{2-}$ compounds, and in consequence have a large quadrupole splitting. $[Ru(SCN)_6]^{3-}$ with one unpaired electron is distorted in the same

Table 5.1 Mössbauer parameters for the $^{99}$Ru resonance [21, 22]

| Oxidation state | Configuration | Compound | δ (Ru metal) /(mm s$^{-1}$) | $\frac{1}{2}e^2qQ_e$ /(mm s$^{-1}$) |
|---|---|---|---|---|
| Ru(VIII) | $4d^0$ $S = 0$ | $RuO_4$ | +1.06 | – |
| Ru(VII) | $4d^1$ $S = \frac{1}{2}$ | $KRuO_4$ | +0.82 | 0.37 |
| Ru(VI) | $4d^2$ $S = 1$ | $BaRuO_4.H_2O$ | +0.38 | 0.44 |
| Ru(V) | $4d^3$ $S = \frac{3}{2}$ | $RuF_5$ | +0.15 | 0.56 |
| Ru(IV) | $4d^4$ $S = 1$ | $RuO_2$ | –0.26 | 0.50 |
| | | $K_2[RuCl_6]$ | –0.31 | 0.41 |
| Ru(III) | $4d^5$ $S = \frac{1}{2}$ | $K_3[RuF_6]$ | –0.84 | – |
| | | $K_2[RuCl_5H_2O]$ | –0.71 | 0.32 |
| | | $[Bu^n_4N]_3[Ru(SCN)_6]$ | –0.49 | 0.53 |
| | | $[Ru(NH_3)_6]Cl_3$ | –0.49 | – |
| | | $[Ru(NH_3)_5Cl]Cl_2$ | –0.53 | 0.38 |
| | | $[Ru(bipyridyl)_3](ClO_4)_3$ | –0.54 | – |
| Ru(II) | $4d^6$ $S = 0$ | $[Ru(NH_3)_6]Cl_2$ | –0.92 | – |
| | | $[Ru(NH_3)_5N_2]Br_2$ | –0.80 | 0.26 |
| | | $[Ru(NH_3)_5NO]Cl_3.H_2O$ | –0.19 | 0.39 |
| | | $K_2[RuCl_5NO]$ | –0.36 | 0.18 |
| | | $K_4[Ru(CN)_6]$ | –0.22 | – |
| | | $K_2[Ru(CN)_5NO].2H_2O$ | –0.08 | 0.39 |
| | | $Ru(CO)_2Cl_2$ | –0.23 | – |
| | | $Ru(\pi\text{-}C_5H_5)_2$ | –0.75 | 0.43 |

way as $[Fe(CN)_6]^{3-}$, and $[RuCl_6]^{2-}$ with two unpaired electrons is similar. The substituted Ru(II) derivatives $[Ru(NH_3)_5NO]^{3+}$ and $[Ru(CN)_5NO]^{2-}$ both show the splitting expected from their non-cubic symmetry. Quantitative analysis of the $^{99}$Ru quadrupole splittings is difficult as yet, particularly in paramagnetic compounds where the large spin-orbit coupling parameter ($\lambda = -1000$ cm$^{-1}$) causes substantial state-mixing and consequent deviations from the value of the quadrupole splitting expected for a particular $4d$-configuration.

It is noteworthy that magnetic ordering has not been detected in any ruthenium co-ordination compound even at 4.2 K, although such ordering has been seen in ruthenium oxide systems at much higher temperatures.

The 73-keV transition in $^{193}$Ir is from the $I_e = \frac{1}{2}$ excited state to the $I_g = \frac{3}{2}$ ground state. The quadrupole spectra are therefore simple doublets, but a large E2/M1 mixing ratio results in all eight magnetic transitions having a finite intensity. Some typical parameters for iridium compounds are given in Table 5.2 [23], and show how the chemical isomer shift decreases with an increase in the number of 5*d* non-bonding electrons in typical 'ionic' halogen compounds. This is analogous to ruthenium and iron. The anomalous shifts of $Na_3[Ir(NO_2)_4Cl_2]$ and $K_3[Ir(CN)_6]$ parallel experience with $K_4[Ru(CN)_6]$ and result from the strong back-donation to $\pi^*$ orbitals.

Although the Ir(III) compounds are diamagnetic because of the $t_{2g}^6$ configuration, several formally octahedral $[IrX_6]^{3-}$ compounds show a substantial quadrupole splitting, implying considerable distortion from regular geometry in the solid state. Iridium(IV) compounds have one unpaired electron, and $K_2IrCl_6$ and $(NH_4)_2IrCl_6$ for example order antiferromagnetically below 3.08 K and 2.16 K respectively. The Fermi contact term, $B_S$, is large at about 50 T per unpaired electron.

The 77.34-keV transition in $^{197}Au$ is from the $I_e = \frac{1}{2}$ excited state to the $I_g = \frac{3}{2}$ ground state and here again the quadrupole spectrum is a doublet. The two common oxidation states of gold are Au(I) and Au(III). The $Au^+$ ion has the closed-shell $5d^{10}$ configuration. Its structural chemistry is dominated by a tendency

Table 5.2 Mössbauer data for the 73-keV transition in $^{193}Ir$

| Oxidation state | Configuration | Compound | $T$ /K | $\delta$† /(mm s$^{-1}$) | $\Delta$‡ /(mm s$^{-1}$) | $B$ /T |
|---|---|---|---|---|---|---|
| Ir(VI) | $5d^3$ $S = \frac{3}{2}$ | $IrF_6$ | 4.2 | +1.45 | −0.47* | 187.6 |
| Ir(V) | $5d^4$ $S = 1$ | $IrF_5$ | 4.2 | +0.06 | 0.74 | – |
| Ir(IV) | $5d^5$ $S = \frac{1}{2}$ | $K_2[IrCl_6]$ | 1.52 | −0.95 | ~0* | 39.5 |
| | | $(NH_4)_2[IrCl_6]$ | 1.42 | −0.99 | ~0* | 40.8 |
| | | $K_2[IrBr_6]$ | 4.2 | −1.10 | −0.20* | 39.8 |
| Ir(III) | $5d^6$ $S = 0$ | $K_3[IrCl_6]$ | 4.2 | −2.26 | 0.52 | – |
| | | $K_3[IrBr_6]$ | 4.2 | −2.23 | 0.40 | – |
| | | $Na_2[IrCl_5H_2O]$ | 4.2 | −1.97 | 0.54 | – |
| | | $Na_3[Ir(NO_2)_4Cl_2]$ | 4.2 | −1.27 | 2.00 | – |
| | | $K_3[Ir(CN)_6]$ | 4.2 | +0.26 | 0.30 | – |

† relative to Ir metal
‡ $\frac{1}{2}e^2qQ_g(1 + \frac{1}{3}\eta^2)^{1/2}$
* from perturbation of the magnetic spectrum.

to form linear compounds of the type $AuX_2^-$ or LAuX. If the bonding is mainly by σ-donation to the gold, the orbitals involved will be 6*s* and $6p_z$, although there is also the possibility of some $5d_{z^2}$ admixture into the 6*s*-orbital.

The $Au^{3+}$ ion has a nominal $5d^8$ configuration and forms square-planar complexes of the type $AuX_4^-$, which for σ-bonding involves the 6*s*, $6p_x$, $6p_y$ and $5d_{x^2-y^2}$ orbitals. Molecular orbital calculations for $AuCl_4^-$ have indicated an effective 0.9 $e^-$ hole in the $5d_{x^2-y^2}$ orbital. In comparing the anions $AuX_2^-$ and $AuX_4^-$ where X is electron withdrawing, the reduced 5*d*-occupation in $AuX_4^-$ would be expected to give a corresponding increase in the *s*-electron density at the gold nucleus as a result of reduced 5*d*-shielding. $\delta R/R$ is believed to be positive for $^{197}Au$, so that $AuX_4^-$ should have a more positive chemical isomer shift than $AuX_2^-$, and as may be seen in Fig.5.10, this is in fact observed. In this plot of some Mössbauer parameters for typical compounds [24], $KAu(CN)_4$ and $AuCl_3$ have larger shifts than in $KAu(CN)_2$ and AuCl respectively. However, these differences are much less than the overall range of shift values for a given oxidation state.

Although neither the shift nor the quadrupole splitting in isolation give a reliable indication of oxidation state, both parameters in conjunction can be used diagnostically. For example $Cs_2Au_2Cl_6$ shows lines from the $AuCl_2^-$ and $AuCl_4^-$ ions and does not contain gold(II).

The isotope shifts observed in the atomic spectrum of mercury have shown that the shielding effects of a 5*d*-electron or 6*p*-electron on the 6*s*-electron density are only 25% and 15% respectively, and the large variation in chemical isomer shift in gold can therefore be attributed mainly to the 6*s*-orbital occupation. If the linear X-Au-X unit is 6*s*6*p* hybridized, then the 6*s*-electron density should increase linearly with the $6p_z$ occupation. There should therefore be a linear relationship in gold(I) compounds between the chemical isomer shift (from the 6*s*-density) and the quadrupole splitting (from the asymmetric 6*p*-occupation). This is found to be the case for those compounds with symmetrical co-ordination shown in Fig.5.10 (AuCl, AuBr, AuI, $Na_3Au(S_2O_3)_2.2H_2O$, $KAu(CN)_2$), and also for AuCN with one bond to carbon and one to nitrogen. Other available data (not shown) for compounds not characterized structurally but believed to possess linear bonding lie close to this line, and there is a good correlation between the observed parame-

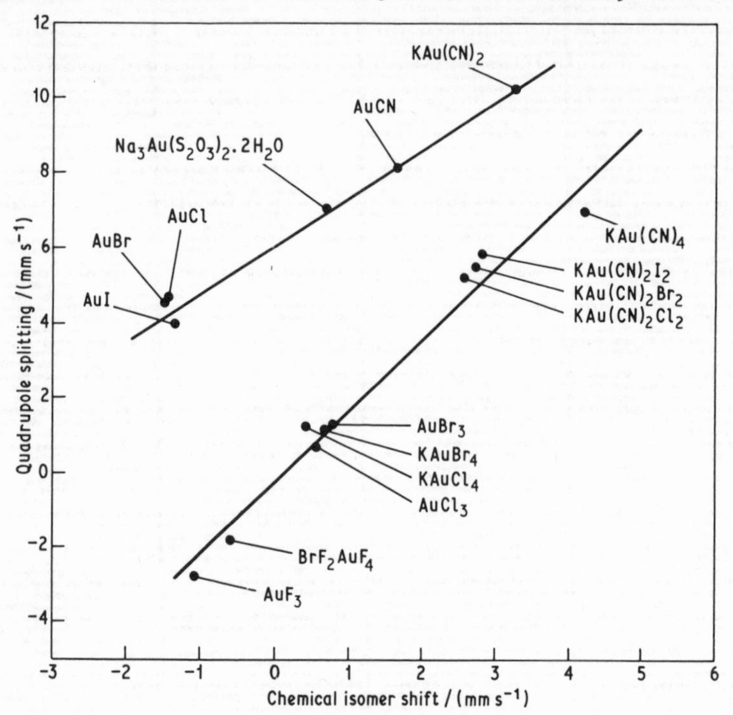


Fig. 5.10 The chemical isomer shifts and quadrupole splittings of some simple gold compounds of the type AuX, $AuX_2^-$, $AuX_3$, $AuX_4^-$ and $AuX_2Y_2^-$. Note the linear correlations shown by gold(I) and gold(III) compounds.

ters and the positions of the ligands in the spectrochemical series.

It is noteworthy that the cyanides correlate with the other compounds, thereby confirming the high $\sigma$-donation of the cyanide group with only weak $\pi$ back-donation in these compounds. The observed linear relationship argues against a varying amount of $5d_{z^2}$ admixture. The value of $q$ in $e^2qQ$ is negative for the $6p_z$ orbital, and since $Q$ is positive, $e^2qQ$ should be negative for all the gold(I) compounds.

In the square-planar gold(III) complexes, the large reduction in the $5d$-density in the $xy$ plane when the gold-ligand bonding is ionic will result in an excess of $5d$-density in the $z$ axis and a negative value of $e^2qQ$. Increasing covalency in the $dsp^2$ bonding will increase the electron density in the $xy$ plane. The effect of the excess of $5d$-density in the $z$ axis will be reduced both directly and by the positive contribution from the $p_x$ and $p_y$ orbitals. The electric field gradient from a $5d$-electron might be anticipated to be greater than from a $6p$-electron because it is in a lower quantum shell with a different $\langle r^{-3} \rangle$ value. However, the large values of $e^2qQ$ found for gold(I) compounds suggests that this is not the case, and that it is the $6p$-electrons which dominate. Increasing covalency might therefore be expected to cause a reversal in the sign of $e^2qQ$, and as seen in Fig.5.9 the correlated data for the gold(III) compounds seem to indicate that this has indeed taken place. The more ionic fluorides may be assumed to have a negative value of $e^2qQ$, whereas the other compounds have the opposite sign. As with the gold(I) cyanides, the gold(III) cyanides show only weak $\pi$-bonding. The isomer shift for $KAu(CN)_2Br_2$ is close to the arithmetic mean for $KAuBr_4$ and $KAu(CN)_4$, implying that the Au-ligand bonds are strongly directional and largely independent of each other.

## 5.6 Lanthanides and actinides

At least one Mössbauer resonance has been observed in each lanthanide element with the exception of cerium, there being no less than 44 resonances in the fourteen elements including lanthanum. Many of the isotopes have a deformed nucleus with a large quadrupole moment, and there has been considerable interest in determining the nuclear parameters of the excited states. From a more chemical point of view, the lanthanide resonances unfortunately provide less useful information. This is partly because the chemistry of each element is dominated by the ionic $M^{3+}$ cation, and the $4f$-electrons are not significantly involved in the bonding.

A typical example is the 21.6-keV resonance of $^{151}Eu$ for which a substantial quantity of data is now available. The observed chemical isomer shifts (quoted relative to $EuF_3$) may be divided into three distinct groups:-

| | |
|---|---|
| ionic $Eu^{2+}$ compounds ($4f^7$) | $-13.9$ to $-10.9$ mm s$^{-1}$ |
| metallic band systems | $-11.4$ to $-7.6$ mm s$^{-1}$ |
| ionic $Eu^{3+}$ compounds ($4f^6$) | $-0.2$ to $+0.9$ mm s$^{-1}$ |

The change in electron configuration from $4f^7$ to $4f^6$ leads to an increase in the $s$-electron density at the nucleus and, because $\delta R/R$

is positive, a net increase in the chemical isomer shift of ~13 mm $s^{-1}$. In any series of related compounds the chemical isomer shift increases as the bonds become less ionic and the degree of 6*s*-character in the bonding increases [25]. For example in the series $EuF_3$, $EuCl_3$, $EuBr_3$ and $EuI_3$ the shift increases by 0.53 mm $s^{-1}$ and in EuOF, EuOCl, EuOBr and EuOI, it increases by 0.74 mm $s^{-1}$. However, in the series EuO, EuS, EuSe and EuTe the shift *decreases* by 1.00 mm $s^{-1}$ because in these compounds the 5*p*-electrons participate in the bonding and alter the screening of the *s*-electrons. The alloys correspond approximately to an $Eu^{2+}$ configuration, but with an increased shift due to 6*s*-participation in the band structure.

The $Eu^{2+}$ ion has a $4f^7$ half-filled shell with an electronic ground state of $^8S_{7/2}$, while the $Eu^{3+}$ ion has a $4f^6$ configuration with a $^7F_0$ ground state which has zero total angular momentum. In neither case is their a valence-electron contribution to the electric field gradient from the 4*f*-electrons. Consequently any quadrupole splitting can only arise from external lattice effects, and being generally less than the linewidth is seen only as line broadening. Thus it is not possible to obtain information easily about the co-ordination or site symmetry.

The $^7F_0$ ground state of $Eu^{3+}$ is non-magnetic, but exchange and crystal-field interactions cause mixing with the $^7F_J$ excited states, and magnetic hyperfine splitting of the $I_g = \frac{5}{2} \rightarrow I_e = \frac{7}{2}$ transition is not uncommon in $Eu^{3+}$ oxides (see Chapter 7). Magnetic splitting has also been observed at very low temperatures in a number of $Eu^{2+}$ compounds such as $EuSO_4$, $EuCl_2$ and $EuCO_3$ [26].

The Mössbauer spectra of the other lanthanides may be interpreted in a similar manner. One major difference is that in some of them (e.g. Sm, Dy, Er and Yb) the electronic ground state is magnetic and the paramagnetic relaxation time may be long enough to allow a magnetic hyperfine splitting at low temperatures. This is considered in more detail in Chapter 6. The values of the magnetic flux density and quadrupole interaction are determined largely by the electronic ground state, and although small variations are observed due to crystal field interactions with the ligands, insufficient data are available to interpret the latter in detail.

Although Mössbauer work with the actinides involves some difficult experimental problems, outstanding results have been obtained for the 59.5-keV $I_g = \frac{5}{2} \rightarrow I_e = \frac{5}{2}$ resonance of $^{237}Np$. The commonly used parent is $^{241}$Am which $\alpha$-decays to $^{237}$Np. The Heisenberg linewidth is only 0.067 mm $s^{-1}$, but the narrowest experimental linewidths are at least 15 times broader. This is probably caused by radiation damage in the source because the $\alpha$-decay displaces the Np nucleus from its original lattice site.

Neptunium forms compounds with oxidation states from $Np^{3+}$ to $Np^{7+}$. The chemical isomer shifts which span a huge range of about 100 mm $s^{-1}$ correlate well with the expected $5f$ configurations of $5f^4$ to $5f^0$. This is illustrated in Fig.5.11 [27]. The majo-

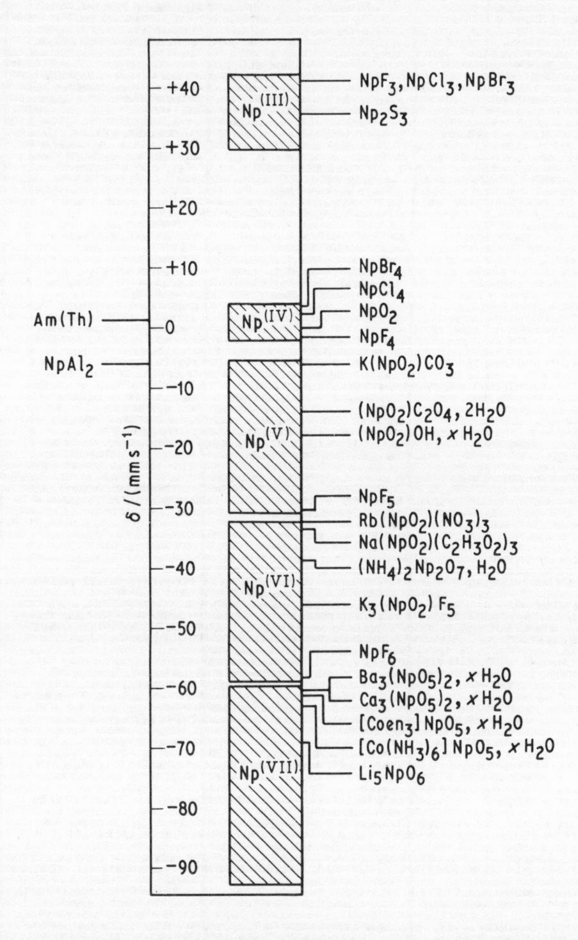


Fig. 5.11 A correlation of the $^{237}$Np chemical isomer shifts with oxidation state. The shaded areas represent the predicted ranges. The more positive shifts in those compounds with co-ordination to oxygen rather than fluorine are indicative of considerable covalent character in the former. (after [27]).

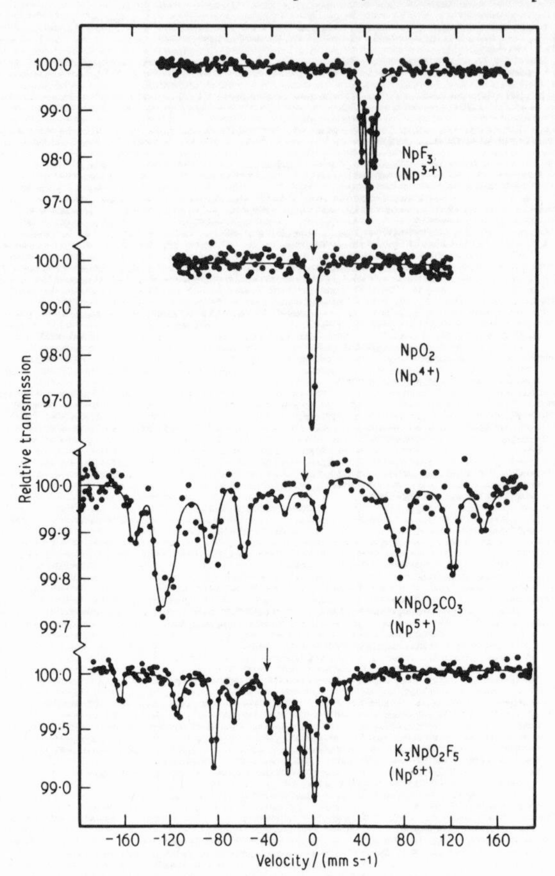


Fig. 5.12 The 59.5-keV $^{237}$Np spectra of four neptunium compounds with different oxidation states showing the large chemical isomer shifts, quadrupole splitting ($NpF_3$), and combined quadrupole/magnetic splitting ($KNpO_2CO_3$ and $K_3NpO_2F_5$). ([28], Fig.4).

rity of spectra have been recorded at 4.2 K, and at this temperature many compounds show a paramagnetic hyperfine splitting due to slow relaxation analogous to the behaviour found in the lanthanides. Quadrupole splittings are also large. Typical spectra [28] are shown in Fig.5.12.

Compared with the lanthanides, the greater variation in chemical isomer shifts within a given oxidation state is a reflection of the greater covalency in these compounds, although the way in which the 5*f*, 6*d* and 7*s* orbitals are utilized is not yet fully understood.

The more positive shifts associated with the $NpO_2^+$ and $NpO_2^{2+}$ cations (in comparison to the fluorides) reflect this increase in covalency. Of particular interest is the compound $Li_5NpO_6$ [27]. It was originally thought to be analogous to $Li_5ReO_6$ in that the $NpO_6^{5-}$ anion has $O_h$ symmetry. However, the Mössbauer spectrum shows a large quadrupole splitting with $\eta = 0.33$, and the neptunium environment is therefore non-cubic. On this basis it is proposed that the octahedron of oxygens is compressed along the $z$ axis and rhombically distorted in the $xy$ plane to emphasize the O-Np-O linear grouping which is so stable in the $NpO_2^+$ and $NpO_2^{2+}$ cations.

## References


[1] Ingalls, R. (1964) *Phys. Rev.*, **133**, A787.
[2] Gibb, T. C. (1968) *J. Chem. Soc. (A)*, 1439.
[3] Ingalls, R., Ono, K. and Chandler, L. (1968) *Phys. Rev.*, **172**, 295.
[4] Nozik, A. J. and Kaplan, M. (1967) *Phys. Rev.*, **159**, 273.
[5] Campbell, M. J. M. (1972) *Chem. Phys. Letters*, **15**, 53.
[6] Dale, B. W. (1971) *J. Phys. C*, **4**, 2705.
[7] Edwards, P. R., Johnson, C. E. and Williams, R. J. P. (1967) *J. Chem. Phys.*, **47**, 2074.
[8] Watson, R. E. and Freeman, A. J. (1961) *Phys. Rev.*, **123**, 2027.
[9] Wertheim, G. K., Guggenheim, H. J. and Buchanan, D. N. E. (1968) *Phys. Rev.*, **169**, 465.
[10] Wertheim, G. K. and Buchanan, D. N. E. (1967) *Phys. Rev.*, **161**, 478.
[11] Johnson, C. E. (1967) *Proc. Phys. Soc.*, **92**, 748.
[12] Johnson, C. E., Rickards, R. and Hill, H. A. O. (1969) *J. Chem. Phys.*, **50**, 2594.
[13] König, E. and Madeja, K. (1967) *Inorg. Chem.*, **6**, 48.
[14] Jesson, J. P., Weiher, J. F. and Trofimenko, S. (1968) *J. Chem. Phys.*, **48**, 2058.
[15] Rickards, R., Johnson, C. E. and Hill, H. A. O. (1968) *J. Chem. Phys.*, **48**, 5231.
[16] Cox, M., Darken, J., Fitzsimmons, B. W., Smith, A. W., Larkworth, L. F. and Rogers, K. A. (1972) *J.C.S. Dalton*, 1192.
[17] Drickamer, H. G. and Frank, C. W. (1973) *Electronic Transitions and the High Pressure Chemistry and Physics of Solids*, Chapman and Hall, London.
[18] Simanek, E. and Wong, A. Y. C. (1968) *Phys. Rev.*, **160**, 348.
[19] Bargeron, C. B., Avinor, M. and Drickamer, H. G. (1971) *Inorg. Chem.*, **10**, 1338.
[20] Fisher, D. C. and Drickamer, H. G. (1971) *J. Chem. Phys.*, **54**, 4825.
[21] Kaindl, G., Potzel, W., Wagner, F., Zahn, U. and Mössbauer, R. L. (1969) *Z. Physik*, **226**, 103.

[22] Potzel, W., Wagner, F. E., Zahn, U. and Mössbauer, R. L. (1970) *Z. Physik,* **240**, 306.
[23] Wagner, F. and Zahn, U. (1970) *Z. Physik,* **233**, 1.
[24] Faltens, M. O. and Shirley, D. A. (1970) *J. Chem. Phys.,* **53**, 4249.
[25] Gerth, G., Kienle, P. and Luchner, K. (1968) *Phys. Letters,* **27A**, 577.
[26] Kalvius, G. M., Shenoy, G. K., Ehnholm, G. J., Katila, T. E., Lounasmaa, O. V. and Reivari, P. (1969) *Phys. Rev.,* **187**, 1503.
[27] Fröhlich, K., Gütlich, P. and Keller, C. (1972) *J.C.S. Dalton,* 971.
[28] Dunlap, B. D., Kalvius, G. M., Ruby, S. L., Brodsky, M. B. and Cohen, D. (1968) *Phys. Rev.,* **171**, 316.

CHAPTER SIX

# Dynamic Effects

Although it is appropriate in many instances to consider the resonant nucleus as being in a static environment, there are several dynamic processes which have important effects on the Mössbauer spectrum. Firstly, the nucleus is vibrating on its lattice site, and the mode of these vibrations depends on the nature of the chemical bonding. Secondly, the nucleus may jump to a different lattice site by some kind of diffusion process, and thirdly there may be large fluctuations in the electronic environment of the nucleus due to relaxation. These will be considered in turn.

## 6.1 Second-order Doppler shift and recoilless fraction

In describing the dynamic motion of the nucleus at its lattice site, it should be noted that the atomic vibrations take place on a time-scale faster than the lifetime of the Mössbauer excited state. Accordingly, the mean velocity $\langle v \rangle$ and the mean displacement $\langle x \rangle$ are both effectively zero. However, the corresponding mean-square parameters are both finite. It was shown in equation (2.8) that the mean-square velocity $\langle v^2 \rangle$ is related to the second-order Doppler shift by

$$\frac{\delta E}{E_\gamma} = -\frac{\langle v^2 \rangle}{2c^2} \tag{6.1}$$

Similarly from equation 1.9, the mean-square displacement $\langle x_j^2 \rangle$

in the direction of observation $j$ is related to the recoilless fraction by

$$f = \exp\left\{-\frac{E_\gamma^2\langle x_j^2\rangle}{(\hbar c)^2}\right\} \tag{6.2}$$

Note that $\langle x_j^2\rangle$ is the mean value along direction $j$, so that it is dependent on direction, whereas $\langle v^2\rangle$ is averaged in all directions and therefore can have only a single value.

Unfortunately the values of $\langle v^2\rangle$ and $\langle x_j^2\rangle$ at any given temperature are a complex function of the dynamic motion of the atom. However, the temperature dependence of either quantity can be compared with the predictions of a suitable theoretical analysis. It is customary to consider the dynamics using the harmonic-oscillator approximation. One such treatment by Housley and Hess [1] calculates the mean-square velocity and mean-square displacement of the particle in direction $j$ to be given by

$$\langle v_j^2\rangle = \frac{\hbar}{m}\sum_i\left[\frac{1}{2} + \frac{1}{e^{\hbar w_i/kT} - 1}\right] b_{ji}^2\omega_i \tag{6.3}$$

$$\langle x_j^2\rangle = \frac{\hbar}{m}\sum_i\left[\frac{1}{2} + \frac{1}{e^{\hbar w_i/kT} - 1}\right] \frac{b_{ji}^2}{\omega_i} \tag{6.4}$$

where $m$ is the mass of the particle, $T$ is the temperature, $k$ is Boltzmann's constant, $\omega_i$ is the frequency of the $i$'th normal vibration mode, $2\pi\hbar = h$ is Planck's constant, and the $b_{ji}$ are constants related to the force constants connecting the particle with its neighbours.

Both these functions have interesting general properties which are illustrated schematically in Fig.6.1. Note that in each case the slope and curvature are zero at $T = 0$, and that the curvature is always greater or equal to zero. The slope approaches a limiting value at high temperatures. If a weighted mean frequency $\omega_j(n)$ is defined as

$$\omega_j(n) = \left[\sum_i b_{ji}^2\omega_i^n\right]^{1/n} \tag{6.5}$$

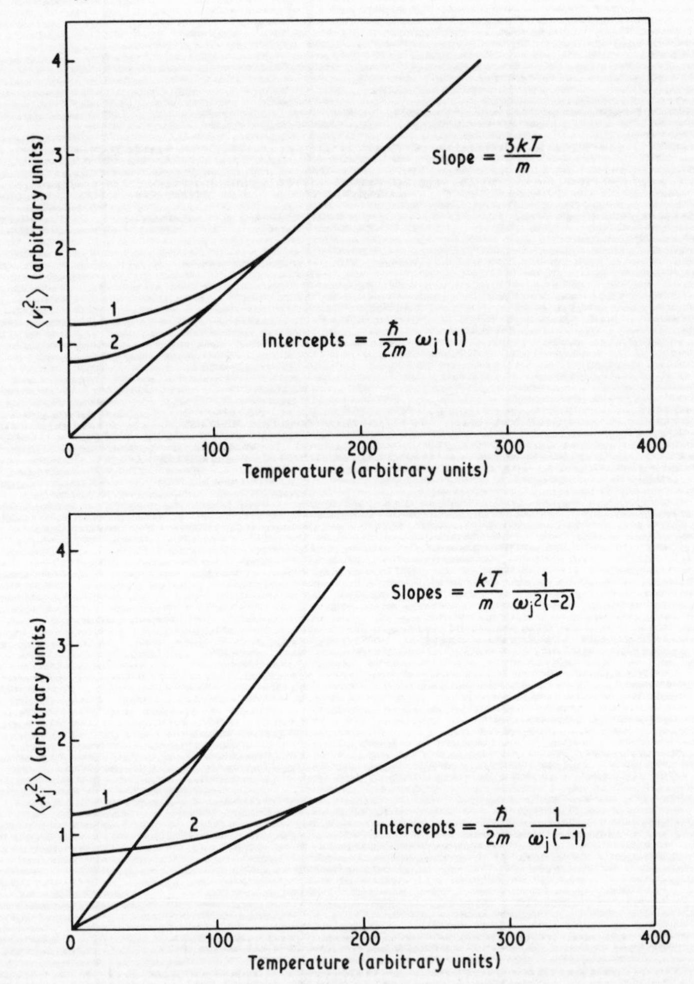


Fig. 6.1 Schematic representations of the temperature dependence of $\langle v_j^2 \rangle$ and $\langle x_j^2 \rangle$. In the former case the intercept is frequency dependent and the limiting slope is frequency independent. In the latter case both parameters are frequency dependent. Two feasible curves for each are labelled 1 and 2. ([1], Fig. 1).

and an expansion is made which is valid for high temperatures, one obtains the limiting conditions

$$\langle v_j^2 \rangle_{T \to \infty} = \frac{kT}{m} \tag{6.6}$$

$$\langle x_j^2 \rangle_{T\to\infty} = \frac{kT}{m} \frac{1}{\omega_j^2(-2)} \tag{6.7}$$

The form of equation 6.6 is particularly interesting as it is independent of frequency. Summing over all three orthogonal directions $j$ to obtain $\langle v^2 \rangle$ (that is $\langle v^2 \rangle = \langle v_x^2 \rangle + \langle v_y^2 \rangle + \langle v_z^2 \rangle$) gives

$$\langle v^2 \rangle_{T\to\infty} = \frac{3kT}{m} \tag{6.8}$$

Substitution in equation (6.1) shows that the second-order Doppler shift reaches a limiting value at high temperature of

$$\left(\frac{\delta E}{E_\gamma}\right)_{T\to\infty} = -\frac{3kT}{2mc^2} \tag{6.9}$$

In the low-temperature limit

$$\langle v_j^2 \rangle_{T\to 0} = \frac{\hbar}{2m}\,\omega_j(1) \tag{6.10}$$

$$\langle x_j^2 \rangle_{T\to 0} = \frac{\hbar}{2m} \frac{1}{\omega_j(-1)} \tag{6.11}$$

Note that these values correspond to a zero-point motion and, being frequency-dependent, will differ from compound to compound.

The evaluation of the general expressions in equation 6.3 and 6.4 requires a detailed knowledge of the frequency distribution which is not usually available, and has therefore to be approximated. The simplest approach is to use the Einstein model with one vibration frequency $\omega_E$ such that the Einstein temperature is $\theta_E = \hbar\omega_E/k$. This then leads to

$$\langle v^2 \rangle = \frac{3k\theta_E}{m}\left[\frac{1}{2} + \frac{1}{e^{\theta_E/T} - 1}\right] \tag{6.12}$$

$$\langle x_j^2 \rangle = \frac{\hbar^2}{km\theta_E}\left[\tfrac{1}{2} + \frac{1}{e^{\theta_E/T} - 1}\right] \tag{6.13}$$

Such an expression for $\langle v^2 \rangle$ has been used for example to fit the second-order Doppler shift of $FeCl_2$, the resulting value being $\theta_E = 169$ K [2].

Better agreement with experiment can usually be obtained using a Debye model in which there is a distribution of frequencies. The number of oscillations with frequency $\omega$ is $N(\omega)$, given by $N(\omega) = 3\omega^2/\omega_0^3$ where $\omega_0$ is the maximum value of $\omega$, and $\hbar\omega_0/k = \theta_D$ defines the 'Debye temperature' of the lattice. Thence

$$\langle v^2 \rangle = \frac{9\hbar}{\omega_0^3 m}\int_0^{\omega_0}\left[\tfrac{1}{2} + \frac{1}{e^{\hbar\omega/kT} - 1}\right]\omega^3 d\omega$$

$$= \frac{9k\theta_D}{m}\left[\tfrac{1}{8} + \left(\frac{T}{\theta_D}\right)^4 \int_0^{\theta_D/T}\frac{x^3 dx}{e^x - 1}\right] \tag{6.14}$$

$$\langle x_j^2 \rangle = \frac{3\hbar}{\omega_0^3 m}\int_0^{\omega_0}\left[\tfrac{1}{2} + \frac{1}{e^{\hbar\omega/kT} - 1}\right]\omega\, d\omega$$

$$= \frac{3\hbar^2}{km\theta_D}\left[\tfrac{1}{4} + \left(\frac{T}{\theta_D}\right)^2 \int_0^{\theta_D/T}\frac{x\, dx}{e^x - 1}\right] \tag{6.15}$$

This latter expression is most commonly used to evaluate the temperature dependence of the recoilless fraction. The lattice temperature ($\theta_E$ or $\theta_D$ depending on which model is used) can be determined directly from the temperature dependence of $f$. The limiting values for the zero-point motion are then

$$\langle v^2 \rangle_E = \frac{3k\theta_E}{2m} \qquad \left(\frac{\delta E}{E_\gamma}\right)_E = -\frac{3k\theta_E}{4mc^2} \tag{6.16}$$

or

$$\langle v^2 \rangle_D = \frac{9k\theta_D}{8m} \qquad \left(\frac{\delta E}{E_\gamma}\right)_D = -\frac{9k\theta_D}{16mc^2} \qquad (6.17)$$

The importance of the zero-point motion as a contribution to the second-order Doppler shift was first recognized by Hazony [3] who showed that it was a far from negligible contribution to the total observed chemical isomer shift. His calculations for $K_4Fe(CN)_6$ and $K_4Fe(CN)_6.3H_2O$ revealed that even closely related compounds can show significantly different zero-point motion contributions, and that comparisons of small differences in the chemical isomer shift are meaningless unless suitable correction is made.

Some of the problems in using the simplified theory are illustrated by recent data for the absorbers iron metal, $Na_2Fe(CN)_5NO.2H_2O$, $Na_4Fe(CN)_6.10H_2O$ and $K_4Fe(CN)_6.3H_2O$. The temperature dependence of the absorption intensity and the chemical isomer shift between 78 K and 293 K were used [4] to derive independent pairs of values for both $\theta_E$ and $\theta_D$. In general, $\theta_E$ and $\theta_D$ as derived from the recoilless fraction data are always lower than from the shift data. This is because the value of $\langle x_j^2 \rangle$ in the general expression is weighted by $\omega_i^{-1}$, while $\langle v_j^2 \rangle$ is weighted by $\omega_i$. Thus $\langle x_j^2 \rangle$ is weighted towards lower frequencies than $\langle v_j^2 \rangle$. This effect is significantly greater in the salts than in the metal because of the greater importance of high-frequency 'optical' modes in the salts. Whereas for iron metal the values of $\theta_D = 358 \pm 18$ K from $\langle x_j^2 \rangle$ and $421 \pm 30$ K from $\langle v^2 \rangle$ agree reasonably well, in sodium nitroprusside for example, the corresponding figures are 203 K and 788 K. Clearly neither the Einstein nor the Debye model is adequate for compounds.

From equation (6.7), the high-temperature limiting value for $\langle x_j^2 \rangle$ is linearly proportional to temperature, so that an extrapolation of high-temperature data should pass through the origin as shown in Fig.6.1. As may be seen in Fig.6.2, this is certainly true in hydrated sodium hexacyanoferrate(II). In hydrated potassium hexacyanoferrate(II) however, the extrapolation does not have a zero intercept, and is indicative of considerable anharmonicity in the lattice vibrations. There is a particularly large mean-square displacement at low temperatures. Potassium hexacyanoferrate(II)

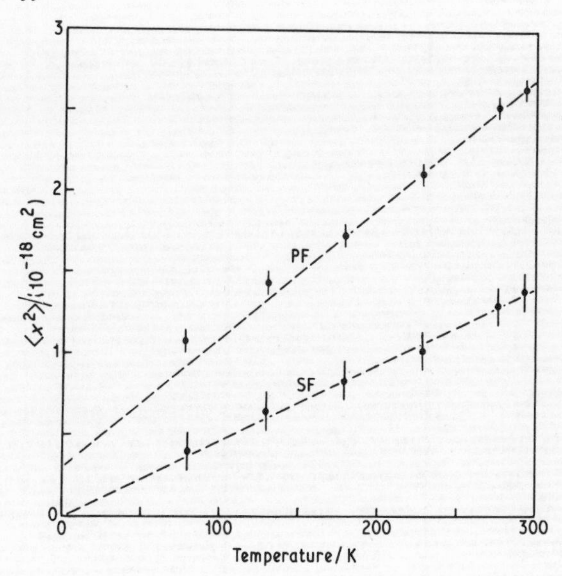


Fig. 6.2 The temperature dependence of $\langle x^2 \rangle$ in hydrated potassium hexacyanoferrate(II), (PF), and hydrated sodium hexacyanoferrate(II), (SF). The large intercept obtained by extrapolation from high temperatures in the former case is indicative of the inadequacy of the harmonic approximation in this compound. ([4], Fig. 3).

becomes ferroelectric below 251 K, and this is known to be often associated with a change in frequency of some of the optical modes of lattice vibration. Low-temperature anharmonicity has also been clearly demonstrated in $FeCl_2$ [2].

Thus far in this discussion of the recoilless fraction, $\langle x_j^2 \rangle$ has been considered to be an isotropic function. This concept is adequate when the absorber is a random microcrystalline sample with a single resonance line and an average value $\langle x^2 \rangle$ can be defined. However, the vibrations of an atom at a lattice site of non-cubic symmetry will in general be anisotropic, and this should result in an anisotropy of the recoilless fraction which should be observable in single-crystal samples. The recoilless fraction in direction $j$ is given by $f = \exp(-\langle x_j^2 \rangle/\lambda\!\!\!^{-2})$ where $\lambda\!\!\!^{-} = \hbar c/E_\gamma$. If we define three principal values in the co-ordinate system $x, y, z$ such that $j$ is defined by the usual polar angles $\theta$ and $\phi$, then

$$\langle x_j^2 \rangle = [\langle x_x^2 \rangle \cos^2 \phi + \langle x_y^2 \rangle \sin^2 \phi] \sin^2 \theta + \langle x_z^2 \rangle \cos^2 \theta \qquad (6.18)$$

In the case of axial symmetry, $\langle x_x^2 \rangle = \langle x_y^2 \rangle$ and $\langle x_j^2 \rangle = \langle x_x^2 \rangle \sin^2 \theta + \langle x_z^2 \rangle \cos^2 \theta$. It is then customary to write $\langle x_z^2 \rangle$ as $\langle x_\parallel^2 \rangle$ and $\langle x_x^2 \rangle$ as $\langle x_\perp^2 \rangle$ so that

$$\langle x_j^2 \rangle = [\langle x_\parallel^2 \rangle - \langle x_\perp^2 \rangle] \cos^2 \theta + \langle x_\perp^2 \rangle \tag{6.19}$$

The angular dependence of $f$ can then be expressed as

$$f(\theta) = \exp \left\{ - \frac{\langle x_\perp^2 \rangle}{\lambda\!\!\!^{-2}} \right\} \exp \left\{ - \frac{\langle x_\parallel^2 \rangle - \langle x_\perp^2 \rangle}{\lambda\!\!\!^{-2}} \cos^2 \theta \right\} \tag{6.20}$$

An example of anisotropic recoilless emission may be found in experiments using a source of $^{57}$Co doped into a single crystal of zinc metal. The crystal was cut at 45° to the $c$ axis so that the recoilless emission can be studied both parallel ($\parallel$) and perpendicular ($\perp$) to the $c$ axis [5]. The results at room temperature of $f_\perp = 0.64$ and $f_\parallel = 0.41$ reflect the large anisotropy in the vibrations of the cobalt impurity atom.

It is more difficult to study anisotropy in single crystal absorbers rather than sources because the observed spectrum has to be corrected for the saturation effects of the finite thickness of the absorber. It was only realized comparatively recently that this requires a rather complicated calculation because of the way in which the absorber polarizes the transmitted $\gamma$-ray beam [6, 7]. One compound in which the vibrational anisotropy at the $^{57}$Fe nucleus has been correctly determined is sodium nitroprusside, $Na_2[Fe(CN)_5NO].2H_2O$ [8]. Both the electric field gradient and mean-square displacement tensors can be represented by three principal values. In the former, only two such values are needed to specify the tensor completely because the sum of all three must be zero, but all three mean-square displacements must be given to specify the vibration completely. Both tensors must satisfy the point symmetry of the lattice site, but in general the two sets of principal axes need not be collinear if the symmetry is low. In sodium nitroprusside, one of the principal values of the mean-square displacement tensor at the $^{57}$Fe site must be parallel to the crystal $c$ axis. The other two are therefore in the $ab$ plane and can be determined from any three measurements within this plane. Full analysis of the experimental data shows that the principal values are directed very close to the crystal axes with $f_c = 0.377$ ($\langle x_c^2 \rangle$

$= 1.83 \times 10^{-18}$ cm²), $f_a = 0.367$ ($\langle x_a^2 \rangle = 1.88 \times 10^{-18}$ cm²) and $f_b = 0.332$ ($\langle x_b^2 \rangle = 2.07 \times 10^{-18}$ cm²). However, although $V_{xx}$ is directed along the *c* axis, $V_{zz}$ is at 36° to the *a* axis in the *ab* plane and is approximately along the N-C-Fe-N-O axis. This is in fact quite reasonable as the electric field gradient will reflect the $C_{4v}$ symmetry of the immediate nearest neighbours, whereas the mean-square displacement will be influenced by long-wavelength vibrations in the crystal reflecting the lattice symmetry, rather than the approximate local symmetry at the resonant atom.

## 6.2 The Goldanskii-Karyagin effect

It was found in Chapter 2 that the quadrupole spectrum of a $\frac{3}{2} \to \frac{1}{2}$ transition in a polycrystalline sample comprises two lines of equal intensity. From Table 2.2, the angular dependence of the $\pm\frac{3}{2} \to \pm\frac{1}{2}$ line in a single crystal is as $(1 + \cos^2\theta)$, and of the $\pm\frac{1}{2} \to \pm\frac{1}{2}$ line is as $(\frac{2}{3} + \sin^2\theta)$ where $\theta$ is the angle between the direction of observation and the principal *z* axis of the electric field gradient. The averaged values for a random polycrystalline sample are then given by integrating over all orientations, and the relative line intensities become

$$\frac{I_{3/2}}{I_{1/2}} = \frac{\int_0^\pi (1 + \cos^2\theta)\sin\theta\,d\theta}{\int_0^\pi (\frac{2}{3} + \sin^2\theta)\sin\theta\,d\theta} = 1 \tag{6.21}$$

These expressions are strictly valid only for axial symmetry, and become more complex for a non-axial field gradient because of state mixing. However, the final result in the more general analysis is still a value of unity.

The above argument is only rigorous if the probability for recoilless absorption (or emission) is the same in all directions. If it is not isotropic, then the intensity of the absorption will be weighted in favour of a particular orientation of the crystallites. Writing the recoilless fraction $f$ as a function of $\theta$, we have for a thin absorber

$$\frac{I_{3/2}}{I_{1/2}} = \frac{\int_0^\pi (1 + \cos^2\theta) f(\theta) \sin\theta \mathrm{d}\theta}{\int_0^\pi (\frac{2}{3} + \sin^2\theta) f(\theta) \sin\theta \mathrm{d}\theta} \neq 1 \tag{6.22}$$

That is to say, an anisotropic recoilless fraction should result in an asymmetry in the intensities of the quadrupole doublet which is independent of the orientation of the sample. This concept is referred to as the Goldanskii-Karyagin effect after its original proponents [9, 10].

That anisotropy of the recoilless fraction should cause unequal line intensities is quite simple to predict, but far harder to verify in practice. This is because there are other phenomena which can cause the same effect. In particular it is extremely difficult to prevent some microcrystallites from orienting on compacting, and of course partial orientation or texture induces asymmetry. The first onset of paramagnetic relaxation (see later) is also seen as an apparent intensity asymmetry which could be mistaken for a Goldanskii-Karyagin effect. Many claimed observations of the effect could well prove to be the result of other causes.

One example which will stand close scrutiny is the behaviour found in $Me_2SnF_2$. The experimental ratio for the line intensities has been measured as $I_{3/2}/I_{1/2} = 0.78$ at 294 K [11]. The X-ray structure of this compound shows a pseudo-octahedral co-ordination about the tin with axial methyl groups and four bridging fluorine atoms. The root-mean-square amplitudes of vibration of the tin as deduced from the X-ray data have been given as 13.7 pm along the Sn-F bonds and 21.0 pm along the Sn-C bonds. The mean-square vibrational amplitude is related to the recoilless fraction by equation (6.20). This expression in terms of the known values for $\langle x_\perp^2 \rangle$ and $\langle x_\parallel^2 \rangle$ given above can be used to evaluate a predicted intensity ratio of 0.72, which is in good agreement with the experimental value.

The principles of the Goldanskii-Karyagin effect can also be applied to magnetic hyperfine spectra and to isotopes with higher spin states; in all cases the effect will be seen as a deviation from predicted line intensities.

## 6.3 Electron hopping and atomic diffusion

In all the dynamic effects discussed so far, the atom or ion has been restricted to a particular site in the lattice. However, there are two situations where this is not strictly true. In the first an electron has a high probability of 'hopping' from the resonant atom (within the Mössbauer lifetime) to another atom on a neighbouring site, and in the second, the resonant nucleus itself changes its lattice site by jump diffusion.

The europium oxide $Eu_3O_4$ has the orthorhombic $CaFe_2O_4$ structure with $Eu^{2+}$ ions at the Ca sites and $Eu^{3+}$ at the Fe sites. These two oxidation states are distinct in the 21.6-keV $^{151}Eu$ Mössbauer spectrum with chemical isomer shifts of $-12.5$ mm $s^{-1}$ ($Eu^{2+}$) and $+0.6$ mm $s^{-1}$ ($Eu^{3+}$) relative to $Eu_2O_3$ [12]. On the other hand, the sulphide $Eu_3S_4$ has a structure in which all the europium cations are in equivalent sites. At low temperatures the $Eu^{2+}$ and $Eu^{3+}$ resonances are distinct with an intensity ratio of 1:2 as expected [13]. Above 210 K the two lines broaden and coalesce until at about 273 K the spectrum comprises a single line at the centre of gravity of the low-temperature spectrum. This behaviour is illustrated in Fig. 6.3. The data can be completely explained by introducing the idea that the extra electron on an $Eu^{2+}$ ion can hop to a neighbouring $Eu^{3+}$ ion by overcoming the energy barrier $E$. All europium sites are involved. This process is characterized by a relaxation time $\tau$ which will decrease with temperature rise because of an enhanced jump probability. The temperature dependence of $\tau$ is given by $\tau = \tau_0 \exp(-E/kT)$. Theoretical calculations of the spectrum predicted for various $\tau$ can be compared with experiment, and were used to derive the relaxation times shown in the figure. As expected, $\tau$ varies exponentially with temperature and leads to an estimate for the activation energy of the hopping process of $E = 22$ kJ $mol^{-1}$.

Hopping processes are also known in iron oxides. $Fe_3O_4$ is an inverse spinel ferrite, $Fe^{3+}[Fe^{2+}Fe^{3+}]O_4$, which shows a transition in many physical properties between 110 and 120 K known as the Verwey transition. Below 110 K, the $Fe^{2+}$ and $Fe^{3+}$ oxidation states are distinct and give magnetic hyperfine splittings which can be separately identified (see Chapter 7 for further details of spinel oxides). Above the transition there is a fast hopping on the B sites which causes substantial changes in the magnetic spectrum as the

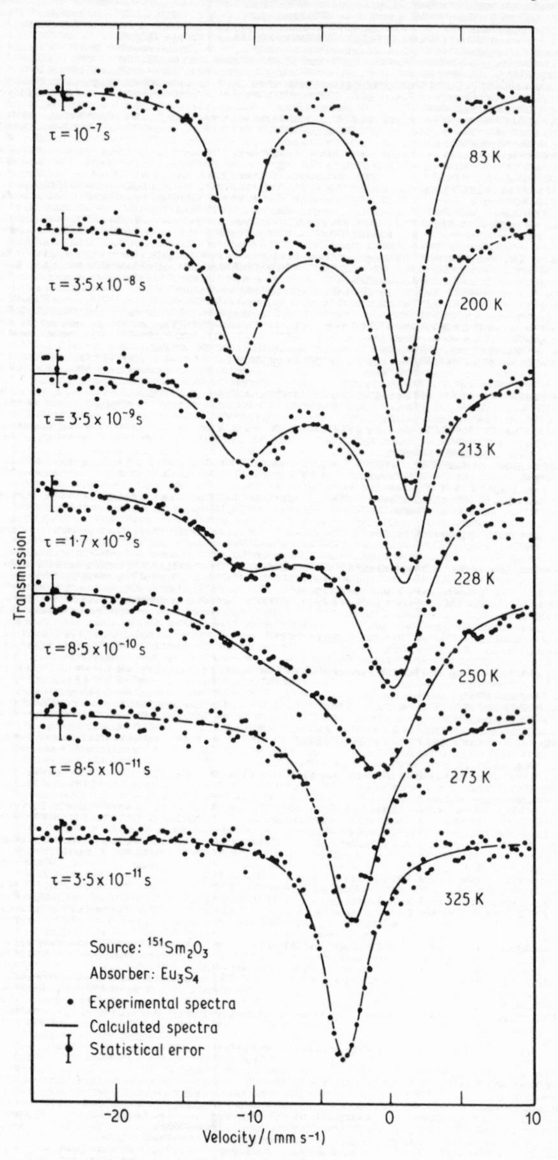


Fig. 6.3 The $^{151}Eu$ spectra of $Eu_3S_4$ between 83 K and 325 K. The weaker line at 85 K is due to $Eu^{2+}$, and the stronger line to $Eu^{3+}$. The fast electron hopping which occurs at higher temperatures results in an averaged spectrum, and the solid curves are computed with a relaxation model whose time constant is $\tau$. ([13], Fig. 1).

$Fe^{2+}$ and 50% of the $Fe^{3+}$ cations combine to give an averaged contribution [14, 15].

The phenomenon of atomic diffusion is more difficult to study. If the nuclei jump instantaneously between lattice sites, then the effect should be manifested as a broadening of the resonance line *without* a change in recoilless fraction. If $\tau_0$ is the mean resting time between jumps, then the fractional increase in the width of the energy distribution at half-height is given by

$$\Delta\Gamma = 2\hbar/\tau_0 \tag{6.23}$$

This can be related to the self-diffusion coefficient $D = l^2/6\tau_0$ where $l^2$ is the mean-square jump distance. Basically similar broadening effects can be expected when the diffusion is a continuous Brownian motion such as is found for instance in very viscous liquids, but this aspect will not be considered further.

Diffusion broadening has been clearly shown for $^{57}Co$ impurity atoms in both copper and gold at high temperatures. However, the broadening observed [16] is less than predicted, and it would seem that any satisfactory theory must include consideration of the probability that a vacancy occurs next to the diffusing ion and is available to accept it. A model incorporating vacancy diffusion has shown that the broadening may be expected to be less than predicted for an uncorrelated jump [17].

Diffusion processes have also been studied in frozen aqueous solutions, using the Mössbauer nucleus as an impurity probe. For example, an aqueous solution containing $Fe^{2+}$ ions forms an unstable phase when quenched to liquid nitrogen temperature. This was formerly thought to be a cubic ice lattice, but more recent evidence favours the formation of a glassy or amorphous variety of ice of near eutectic composition (at least in the region surrounding the impurity ions), so that the freezing causes segregation into two components, a eutectic and a pure ice phase [18]. At about 163 K the glass softens and at about 183 K the supercooled liquid recrystallizes *irreversibly* to a new phase based on the stable hexagonal ice lattice which is then retained until the eutectic melting point. While the transition from the low to the high temperature form is taking place, the $^{57}Fe$ Mössbauer resonance completely disappears, presumably because of the effect of diffusional motion on the recoilless fraction. The $^{57}Fe$ quadrupole splitting has different values above and below the phase transition temperature.

Quenched solutions containing $EuCl_3$ show an entirely analogous behaviour, but for some reason a quenched solution of $Fe^{2+}$ *and* $Eu^{3+}$ is completely different [19]. Both the $^{57}Fe$ and $^{151}Eu$ resonances decrease rapidly above 183 K and vanish at about 208 K presumably due to diffusion, but do not reappear above this temperature. Re-cooling causes a reappearance of the original spectra, showing that in this case the effect is fully *reversible* and that the ice transformation is strongly inhibited in the mixed solution. There is no unique interpretation of these observations at present, but clearly the Mössbauer resonance has a valuable role to play in the study of diffusion in quenched solutions.

## 6.4 Paramagnetic relaxation

During the discussion of paramagnetic complexes in Chapter 5, it was shown that although there is an interaction between the nuclear spin and the electronic spin, the relaxation of the latter is generally faster than the lifetime of the Mössbauer excited state so that the average value $\langle S_z \rangle$ defined in the nuclear axis system is zero. This contrasts with magnetically ordered materials where the long-range coupling induces a non-zero value of $\langle S_z \rangle$ which is seen directly as the Fermi contact interaction.

However, if the electronic relaxation rate in a paramagnetic material is very slow, there will again be a static hyperfine interaction between the nucleus and the spin $S$ (this paramagnetic interaction differs from the magnetically ordered case in which $\langle S_z \rangle$ is still a dynamic average). It is necessary to consider this static case corresponding to a very slow relaxation in some detail, before turning to the intermediate dynamic situation. The static spin $S$ may be able to couple in several ways with the nucleus. For example the energy levels of the $Fe^{3+}$ ion ($S = \frac{5}{2}$) in an axial ligand field are three Kramers' doublets which are normally represented as $|S_z = \pm\frac{1}{2}\rangle$, $|\pm\frac{3}{2}\rangle$, $|\pm\frac{5}{2}\rangle$. The $Fe^{III}$ ($S = \frac{1}{2}$) ion has a strong spin-orbit interaction, and the 6-fold degenerate $^2T_2$ term is also split into three Kramers' doublets. If we consider a $|\pm\frac{1}{2}\rangle$ Kramers' doublet state, both the $S_z = +\frac{1}{2}$ and $S_z = -\frac{1}{2}$ orientations couple with the nuclear spin to give an observed magnetic flux density of magnitude $B$, but with opposite sign. There is no preferred direction for $S_z$ (unlike the situation in ordered systems) and because the Mössbauer spectrum does not indicate the sign of $B$, we expect to see only a single magnetic hyperfine splitting corresponding to the flux density $|B|$. The three Kramers' doublets of the $Fe^{3+}$ ion are generally separated by only a very small energy. At temperatures above 1 K all three are partly populated and one can in principle expect to see three superimposed hyperfine splittings.

A full mathematical treatment of the interactions is often complex. A comprehensive discussion of the principles involved has been given by Wickmann and Wertheim [20]. A useful representation is to consider each Kramers' doublet as having an effective spin of $S' = \frac{1}{2}$ and accumulating all other terms into a hyperfine tensor $A$ so that the general form of the Hamiltonian operator becomes

$$\mathscr{H}_{\mathrm{M}} = A_x S'_x I_x + A_y S'_y I_y + A_z S'_z I_z \tag{6.24}$$

Considering the $Fe^{3+}$ ion, the $|S_z = \pm\frac{5}{2}>$ doublet corresponds to the uniaxial symmetry $A_z > A_x = A_y \simeq 0$ and reduces the Hamiltonian to $\mathscr{H}_{5/2} = A_z S'_z I_z$. The corresponding spectrum is a simple six-line spectrum similar in appearance to the magnetically ordered pattern. This form of Hamiltonian is also commonly applied in the rare-earth elements. The case $A_x = A_y = 3A_z$ corresponds to the $Fe^{3+}$ $|S_z = \pm\frac{1}{2}>$ doublet which is more complicated with a total of ten resolvable hyperfine components in the spectrum.

In general there may be an additional quadrupole interaction in the Hamiltonian, plus a term representing an externally applied field and written as an isotropic interaction $\mathscr{H}_{\mathrm{E}} = A\,\mathbf{I.S}$. The observed behaviour can vary drastically in external fields having a flux density of only a few tesla. The static situation as described here can rarely be achieved for all the Kramers' doublets involved. Before discussing the spectra to be expected, it is necessary to consider the effects of relaxation.

The electronic spin can change its quantized orientation by three types of relaxation process. In spin-lattice relaxation it can transfer or receive energy from a lattice phonon with a mean-time between transitions denoted by $T_1$. The interaction takes place via the spin-orbit interaction $\lambda\mathbf{L.S}$ so that if the ion has considerable orbital momentum, i.e. $\langle L \rangle$ is large, then the relaxation will be rapid. $Fe^{2+}$ and low-spin $Fe^{\mathrm{III}}$ compounds do not normally show paramagnetic hyperfine effects because the spin-lattice relaxation is fast. On the other hand, $Fe^{3+}$ which is an $S$-state ion has a correspondingly long spin-lattice relaxation time. Because the process involves lattice phonons, this type of relaxation will slow down with decreasing temperature.

Spin-spin relaxation involves a mutual spin-flip between neighbouring ions of the type $S_{1+}S_{2-} \rightarrow S_{1-}S_{2+}$ and is energy conserving if the spins are identical. Because the spins are on neighbouring ions, the relaxation is strongly concentration-dependent and will be much slower as the distance between ions increases. However, unlike spin-lattice relaxation, spin-spin relaxation is not temperature-dependent. The third type of relaxation is less important and involves a cross-relaxation between two different spin types. In the current context this could mean a nuclear spin and an electronic spin.

The result of these various relaxation processes on a particular energy level is usually expressed by a single effective parameter $\tau$, which represents the lifetime of the electronic level. This must be compared with the nuclear Larmor precession frequency $\omega_L$, and if $\tau \gg \omega_L^{-1}$ the static spectrum is seen. When $\tau \simeq \omega_L^{-1}$, the relaxation randomly changes the energy of any particular Mössbauer transition. For example the transition $|S_z = +\frac{1}{2}\rangle \rightarrow |S_z = -\frac{1}{2}\rangle$ reverses the sign of the flux density $B$. The detailed theory is again too complex to give here [20], but the behaviour to be expected can be illustrated by examples.

The rare earth elements are comparatively simple to treat in that a static paramagnetic hyperfine splitting is usually produced by a highly anisotropic hyperfine tensor with $A_z \gg A_x, A_y$. Thus the Hamiltonian reduces to $\mathscr{H} = A_z S'_z I_z$ as already mentioned, and the nuclear level with spin $I$ splits into $2I + 1$ equi-spaced levels as with the normal Zeeman interaction for long-range order. If the anisotropy were reduced, the spin-spin relaxation time would decrease and the hyperfine splitting disappear.

An example is given by the $Dy^{3+}$ ion which has a $^6H_{15/2}$ electronic ground state. The large spin-orbit interaction in the rare-earths causes $J$ to be a good quantum number, and the excited states are not thermally accessible from the $|J_z = \pm\frac{15}{2}\rangle$ ground state doublet level at low temperatures. The $^{161}Dy$ resonance in $DyCrO_3$ at 4.2 K comprises a static paramagnetic hyperfine splitting from the $Dy^{3+}$ cation, and is illustrated in Fig. 6.4. There are a

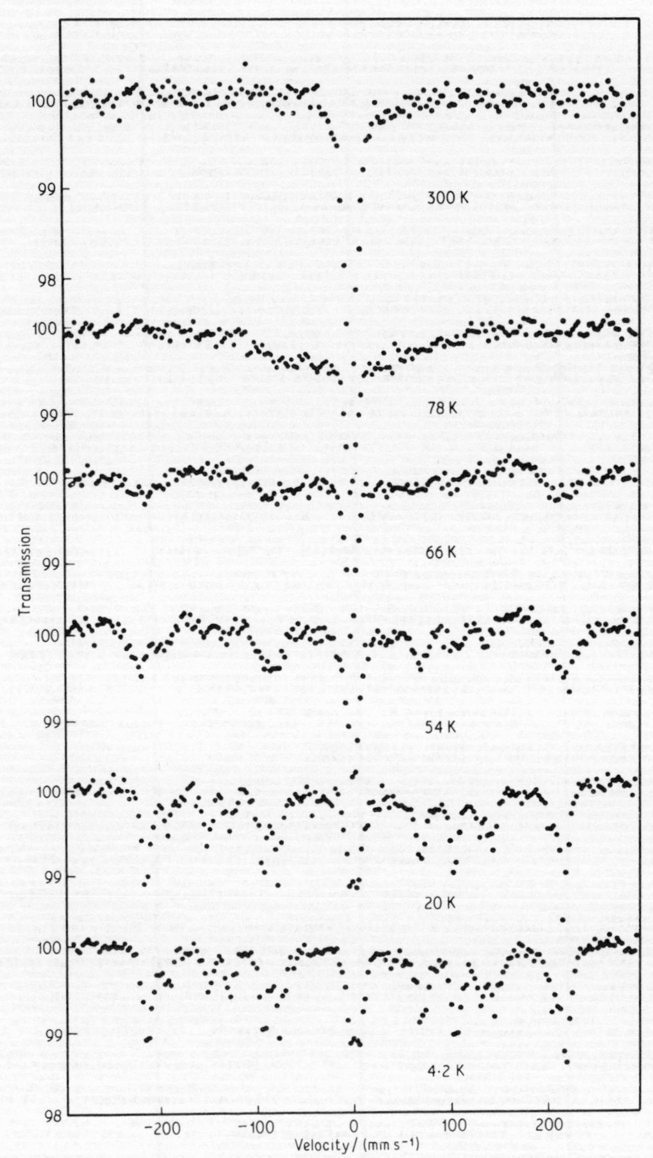


Fig. 6.4 The 25.65-keV $^{161}$Dy resonance in $DyCrO_3$. At 4.2 K the spectrum is a resolved paramagnetic hyperfine splitting from the $|J_z = \pm\frac{15}{2}\rangle$ ground state. Increasing temperature causes a reduction in the spin-lattice relaxation time and a complete collapse of the splitting. ([21], Fig. 1).

total of 16 resonance lines from the $I_e = \frac{5}{2} \rightarrow I_g = \frac{5}{2}$ E1 decay [21]. As the temperature is raised the effective relaxation time decreases, being 7 ns at 20 K and 0.05 ns at 78 K, and results in complete collapse of the magnetic structure. The inward 'motional narrowing' of the spectrum is typical of relaxation processes. The decreasing relaxation time is a result of a faster spin-lattice relaxation as thermal excitation takes place to the excited $J_z$ levels. The static spectrum at 4.2 K is slightly asymmetrical because of an axially symmetric quadrupole splitting. Very similar spectra are known in a large number of dysprosium compounds at 4.2 K including $DyF_3.5H_2O$, $Dy(NO_3)_3.6H_2O$ and $DyPO_4.5H_2O$ [22].

Fully resolved paramagnetic hyperfine splitting has also been recorded for $^{149}Sm$ ($J_z = \pm\frac{5}{2}$) in $SmCl_3.6H_2O$ and $Sm(NO_3)_3.6H_2O$ at 4.2 K [23], and in several of the erbium and ytterbium resonances. Only one case is known of paramagnetic hyperfine structure in a rare-earth compound which is not highly anisotropic. Erbium ethyl sulphate, $Er(EtSO_4)_3.9H_2O$, has short relaxation times, but magnetic dilution with 97.6% of the non-magnetic ytterbium compound slows the spin-spin relaxation rate and allows the appearance of a static field in the 80.6-keV $^{166}Er$ resonance [24]. In this case the abbreviated Hamiltonian is not valid and results in incorrectly calculated line positions.

There are many examples of paramagnetic relaxation in the Mössbauer spectra of iron compounds, the majority of these containing $Fe^{3+}$ ions. This *S*-state ion does not relax by a spin-lattice mechanism, and the spin-spin process dominates. In iron(III) compounds with a small Fe-Fe distance such as $FeCl_3.6H_2O$ (~6 Å = 600 pm) the relaxation is fast and the lines are comparatively sharp, whereas compounds with large chelating ligands such that the Fe-Fe distance increases to about 9.5 Å tend to show definite magnetic structure [25]. The spectrum of iron(III) acetylacetonate (7.6 Å) is very broad at 1.8 K with a non-Lorentzian lineshape of about ten times the natural linewidth and only narrows to about 7 linewidths as the temperature is raised to 300 K.

The relaxation behaviour observed may be particularly complex when the resonance is also quadrupole-split. A common type of behaviour in $^{57}Fe$ spectra is that one of the two lines becomes increasingly broadened with *rise* in temperature. This is the inverse of the normal situation, and is found for example in iron(III) haemin chlorides (see Chapter 11), and in the acetylacetonate complex, $Fe(acac)_2Cl$ [26]. These compounds all feature a square-pyramidal geometry with an apical chlorine atom and are in the high-spin $S = \frac{5}{2}$ configuration rather than the intermediate-spin $S = \frac{3}{2}$ configuration found in the dithiocarbamato derivatives. However, the phenomenon also occurs in octahedrally co-ordinated compounds such as $FeCl_3.6H_2O$.

Typical spectra for $Fe(acac)_2Cl$ are shown in Fig. 6.5. The symmetrical doublet seen at 4.2 K becomes grossly broadened by 300 K, implying that the relaxation slows down as the temperature *increases* which is the opposite of spin-lattice relaxation. An explanation has been provided by Blume [27]. The ground state of the $Fe^{3+}$ ion in the crystal field is the $|\pm\frac{1}{2}\rangle$ state with the $|\pm\frac{3}{2}\rangle$ and $|\pm\frac{5}{2}\rangle$ excited states at energies $\Delta$ and $3\Delta$ above the ground state (Fig. 6.6). The separation $\Delta$ expressed in terms of $kT$ is usually of the order of 10–20 K (80–160 J mol$^{-1}$) so that at 4.2 K only the $|\pm\frac{1}{2}\rangle$ level is occupied. The spin-spin relaxation process involves a $S_{1+}S_{2-} \rightarrow S_{1-}S_{2+}$ transition and the rate is temperature independent *provided* that the spin populations remain constant. In fact, the $|\pm\frac{1}{2}\rangle$ level population decreases substantially as an increase in temperature causes progressive thermal population of the higher levels. The result is an increase in the relaxation time of the $|\pm\frac{1}{2}\rangle$ doublet. The excited states themselves have much longer relaxation times, and the net increase in relaxation time results in the line broadening. It is usually the $I_g = \pm\frac{1}{2} \rightarrow I_e = \pm\frac{3}{2}$ component of the quadrupole splitting which broadens most, so

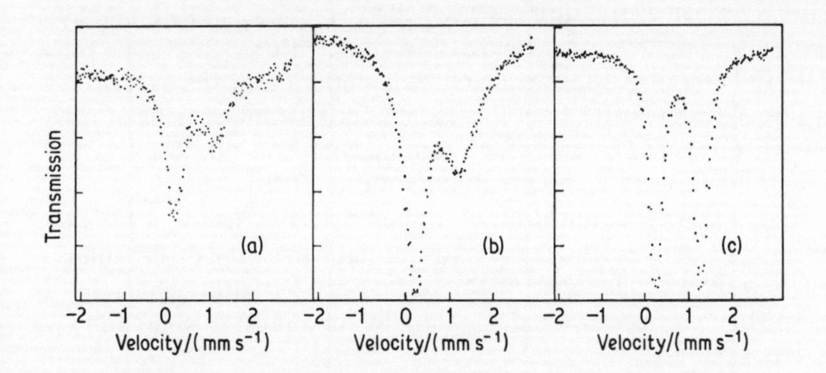


Fig. 6.5 Mössbauer spectra of $Fe(acac)_2Cl$ at (a) room temperature, (b) 80 K and (c) 4.2 K, showing the progressive decrease in asymmetry as the relaxation rate increases. ([26], Fig.1).

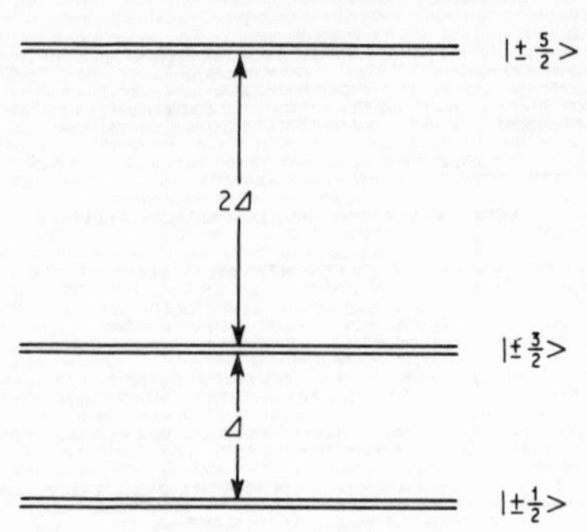


Fig. 6.6 The energy-level scheme of the $Fe^{3+}$ ion.

that $e^2qQ$ is seen to be negative in $FeCl_3.6H_2O$ and positive in the square-pyramidal compounds.

The compounds just discussed are magnetically dilute in that the iron atoms do not strongly interact with each other. It is interesting to note that in compounds where strong intramolecular antiferromagnetic coupling takes place between pairs of $Fe^{3+}$ ions such as $[Fe(salen)]_2O$, $[Fe(salen)Cl]_2$ where salen-1,2-di(salicylideneamino)ethane, and $[(phen)_2Fe.O.Fe(phen)_2]$ where phen = 1,10-phenanthroline, the quadrupole lines are sharp at low temperature because the ground state is non-magnetic with a resultant spin of $S' = 0$ [26]. However, the excited states with $S' = 1, 2, 3,$ 4 and 5 are magnetic and result in similar relaxation broadening to the previous case except that it is not usually as pronounced [28]. A clear distinction between the two cases can be made by applying a large external field at low temperature [29]. The $|\pm\frac{1}{2}\rangle$ state in $Fe(acac)_2Cl$ shows a large augmentation of the external magnetic flux density (e.g. $B = 13.5$ T for an applied field of flux density $B_0 = 3$ T) whereas for a non-magnetic $S' = 0$ state the external flux density is essentially unmodified [28].

In general, the application of a small external field to a broadened $Fe^{3+}$ spectrum often causes substantial narrowing of the lines due to a field dependence of the relaxation, but although such narrowing has been studied in several compounds, it is not fully understood.

It is noteworthy that $Fe^{3+}$ relaxation is not seen in oxides containing more than 5% of $Fe^{3+}$ cations because the short cation-cation distances make them magnetically concentrated, but resolved splitting can be expected below the 2% $Fe^{3+}$ level. The spectra of $Fe^{3+}$ ions (0.14%) in $Al_2O_3$ are typical of the behaviour found, and the $^{57}Fe$ spectrum at 77 K is shown in Fig. 6.7 [30]. Although the ground state is the $|\pm\frac{1}{2}\rangle$ level, the relaxation rate of this level is faster than those of the excited states, and consequently it contributes only a broad background to the spectrum. The best resolved lines are from the $|\pm\frac{5}{2}\rangle$ level with a flux density of about 54.8 T (close to that expected for an $S = \frac{5}{2}$ ion), and it appears that relaxation is also slow for the $|\pm\frac{3}{2}\rangle$ state. Increasing the iron content decreases the spin-spin relaxation

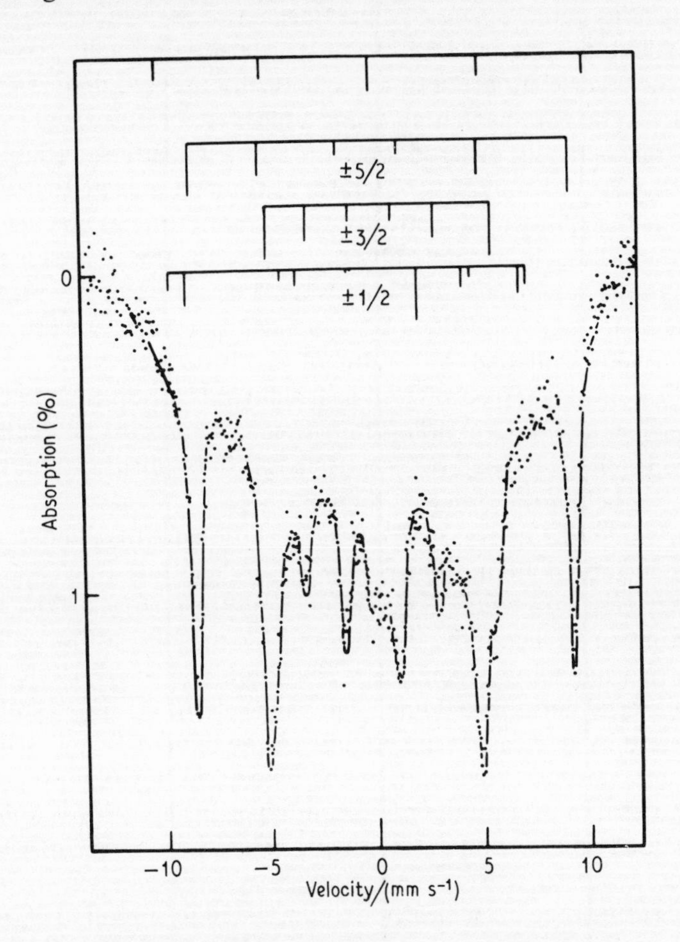


Fig. 6.7 The $^{57}Fe$ spectrum at 77 K of 0.14 at.% $Fe^{3+}$ in $Al_2O_3$. The predicted lines for the three Kramers' doublets are shown as bar diagrams, but the $|\pm\frac{1}{2}\rangle$ contributions are not resolved. ([30], Fig. 2).

times and causes collapse of the magnetic splittings, and this process is largely complete at 1.4% Fe.

Of the other common iron electronic configurations, low-spin iron(III) shows magnetic effects only if magnetically dilute (<5%) to prevent spin-spin relaxation and at low temperature to slow the spin-lattice relaxation. Such conditions have been achieved by doping $K_3Fe(CN)_6$ into $K_3Co(CN)_6$ [31].

## 6.5 Superparamagnetism

Having considered the case of paramagnetic ions which can show magnetic hyperfine interactions, it is appropriate finally to consider the inverse situation of a magnetically ordered material which appears to be paramagnetic. Using a naïve description, one can say that a magnetically ordered ion with $S_z = \frac{1}{2}$ will still fluctuate between the $S_z = +\frac{1}{2}$ and $S_z = -\frac{1}{2}$ states as for a paramagnetic ion, but that the spin will statistically spend more time in one state than the other. In magnetically ordered solids the spin-exchange interactions are strong. The spin-wave description of the magnetism results in an averaged value $\langle S_z \rangle$ which is responsible for the magnetic hyperfine splitting. This value of $\langle S_z \rangle$ is defined along the direction of magnetization and has a long relaxation time because the co-operative interaction is long-range. Thus a magnetically ordered solid usually shows a static hyperfine splitting at all temperatures below the ordering temperature.

The value of $\langle S_z \rangle$ is not, however, a static quantity, and within each magnetic domain fluctuates with a relaxation time $\tau = \tau_0 \exp(KV/kT)$ where $KV$ is effectively the energy barrier to be overcome in changing the direction of the magnetization, and $kT$ is the thermal energy. Since $V$ is the volume of the domain, $K$ is the energy per unit volume. Thus the relaxation time can be decreased by decreasing the volume of the domain or by raising the temperature. Provided that the relaxation rate is slower than the Mössbauer lifetime then the full hyperfine splitting is seen, but if the two are comparable then a relaxation narrowing can be expected.

This effect has been clearly demonstrated in several iron oxides such as $\alpha$-$Fe_2O_3$ [32] and $\alpha$-FeOOH [33] by decreasing the particle size and thus the magnetic domain size to the order of <200 Å (20 nm). Although the magnetic hyperfine splitting is still present at low temperatures, it is seen to collapse with increase in temperature to give the spectrum characteristic of the paramagnetic phase well before the Néel temperature is reached. The finer details of the relaxation are often obscured because most methods of generating small particles tend to give a distribution in particle size with a consequent distribution in the relaxation time.

Probably the most common type of magnetic relaxation occurs in non-stoichiometric compounds where the ordered cations on a particular lattice site have been diluted with a non-magnetic substituent. This weakens the exchange interactions to the point where they become localized to small clusters of cations. It is thus analogous to the small particle-size effect, and a contributory cause of the collapse of the magnetic hyperfine splitting. Several examples will be found in Chapter 7.

## References


[1] Housley, R. M. and Hess, F. (1966) *Phys. Rev.*, **146**, 517.
[2] Johnson, D. P. and Dash, J. G. (1968) *Phys. Rev.*, **172**, 983.
[3] Hazony, Y. (1966) *J. Chem. Phys.*, **45**, 2664.
[4] Lafleur, L. D. and Goodman, C. (1971) *Phys. Rev. B*, **4**, 2915.
[5] Housley, R. M. and Nussbaum, R. H. (1965), *Phys. Rev.*, **138**, A753.
[6] Housley, R. M., Gonser, U. and Grant, R. W. (1968) *Phys. Rev. Letters*, **20**, 1279.
[7] Housley, R. M., Grant, R. W. and Gonser, U. (1969) *Phys. Rev.*, **178**, 514.
[8] Grant, R. W., Housley, R. M. and Gonser, U. (1969) *Phys. Rev.*, **178**, 523.
[9] Goldanskii, V. I., Gorodinskii, G. M., Karyagin, S. V., Korytko, L. A., Krizhanskii, L. M., Makarov, E. F., Suzdalev, I. P. and Khrapov, V. V. (1967) *Doklady Akad. Nauk S.S.S.R.*, **147**, 127.
[10] Karyagin, S. V. (1963) *Doklady Akad. Nauk S.S.S.R.*, **148**, 1102.
[11] Herber, R. H. and Chandra, S. (1970) *J. Chem. Phys.*, **52**, 6045.
[12] Wickman, H. H. and Catalano, E. (1968) *J. Appl. Phys.*, **39**, 1248.
[13] Berkooz, O., Malamud, M. and Shtrikman, S. (1968) *Solid State Comm.*, **6**, 185.
[14] Banerjee, S. K., O'Reilly, W. and Johnson, C. E. (1967) *J. Appl. Phys.*, (1967) **38**, 1289.
[15] Hargrove, R. S. and Kundig, W. (1970) *Solid State Comm.*, **8**, 803.
[16] Knauer, R. C. and Mullen, J. G. (1968) *Appl. Phys. Lett.*, **13**, 150.
[17] Knauer, R. C. (1971) *Phys. Rev. B*, **3**, 567.
[18] Cameron, J. A., Keszthelyi, L., Nagy, G. and Kacsóh, L. (1971) *Chem. Phys. Letters*, **8**, 628.
[19] Dilorenzo, J. V. and Kaplan, M. (1968) *Chem. Phys. Letters*, **2**, 509.

[20] Wickman, H. H. and Wertheim, G. K. (1968) In *Chemical Applications of Mössbauer Spectroscopy,* Chapter 11, ed. V. I. Goldanskii and R. H. Herber, Academic Press, New York.
[21] Eibschutz, M. and van Uitert, L. G. (1969) *Phys. Rev.,* **177**, 502.
[22] Wickman, H. H. and Nowik, I. (1967) *J. Phys. Chem. Solids,* **28**, 2099.
[23] Ofer, S. and Nowik, I. (1967) *Nuclear Phys.,* **A93**, 689.
[24] Seidel, E. R., Kaindl, G., Clauser, M. J. and Mössbauer, R. L. (1967) *Phys. Letters,* **25A**, 328.
[25] Wignall, J. W. G. (1966) *J. Chem. Phys.,* **44**, 2462.
[26] Cox, M., Fitzsimmons, B. W., Smith, A. W., Larkworthy, L. F. and Rogers, K. A. (1969) *Chem. Comm.,* 183
[27] Blume, M. (1967) *Phys. Rev. Letters,* **18**, 305.
[28] Buckley, A. N., Herbert, I. R., Rumbold, B. D., Wilson, G. V. H. and Murray, K. S. (1970) *J. Phys. Chem. Solids,* **31**, 1423.
[29] Fitzsimmons, B. W. and Johnson, C. E. (1970) *Chem. Phys. Letters,* **6**, 267.
[30] Johnson, C. E., Cranshaw, T. E. and Ridout, M. S. (1964) *Proc. Int. Conf. on Magnetism, Nottingham,* (Inst. Phys. Physical Soc., London, 1965), p. 459.
[31] Oosterhuis, W. T. and Lang, G. (1969) *Phys. Rev.,* **178**, 439.
[32] Kundig, W., Bömmel, H., Constabaris, G. and Lindquist, R. H. (1966) *Phys. Rev.,* **142**, 327.
[33] van der Kraan, A. M. and van Loef, J. J. (1966) *Phys. Letters,* **20**, 614.

CHAPTER SEVEN

# *Oxides and Related Systems*

There are a large number of ionic compounds in which the crystal structure is determined largely by a close-packed lattice of the anions. The most familiar examples are found among the oxides, sulphides, and anhydrous halides. The anionic lattice has the power to accommodate a wide range of different cations in the largest interstices, not only to give stoichiometric compounds, but also to form the important category of non-stoichiometric compounds. Any given cation in the lattice is co-ordinated to several anions which frequently number four or six and are arranged in a near regular geometry. The electronic interaction of the cation with this environment can be described in terms of the ideas outlined in the previous chapters on electronic structure. However, although all cations of a given type may have the same co-ordination to the anion, the successive co-ordination spheres containing other cations may well be very different. It is this distinction between a stoichiometric ordered lattice, such as in $FeF_2$, and a non-stoichiometric disordered lattice as in $Co_{1-x}Zn_xFe_2O_4$ which results in important new phenomena.

The examples which follow will be restricted mainly to oxides with the spinel or iron garnet lattices in the knowledge that the principles outlined for these compounds are directly applicable to other structures. It is appropriate in the first instance to develop the concepts introduced in the earlier chapters with particular reference to those oxides and sulphides which are close to stoichiometry, as these serve as model compounds for their more complicated derivatives. Some topics involving dynamic effects such as electron hopping, paramagnetic relaxation and superparamagnetism have been described separately in Chapter 6.

## 7.1 Stoichiometric spinels

One of the most widely adopted oxide structures is the spinel lattice, and because it has both tetrahedral and octahedral co-ordination at the cation sites, it serves as a convenient illustrative example. The idealized lattice is shown in Fig. 7.1, and consists of tetrahedral (A) sites and octahedral [B] sites in a face-centred cubic oxide sublattice, with stoichiometry $AB_2O_4$. Where necessary the convention of round parentheses for the A-site and square brackets for the B-site will be used.

The tetrahedral site has cubic ($T_d$) symmetry, and hence there is no 'lattice' contribution to the electric field gradient, but the octahedral site has trigonal symmetry and one can therefore anticipate a large electric field gradient along the [111] trigonal axis of the octahedron. Although many oxides with a 'normal' $(A)[B_2]O_4$ structure exist, others are of the type $(B)[AB]O_4$ and are said to be 'inverse'. Partial inversion is also common. The Mössbauer spectrum can be used to determine the site occupancy in the spinel (and in other oxides with multiple site symmetries). Many of the spinels can order magnetically. The major magnetic interactions are between A-site and B-site cations, the A-A and B-B interactions being much weaker. In the context of the $^{57}Fe$ resonance, this may involve both Fe-Fe and Fe-M interactions where M is another paramagnetic cation.

Several general observations can be made about the $^{57}Fe$ spectra of spinels which derive from the discussion in Chapters 4 and 5. The chemical isomer shifts increase in the sequence $\delta$(tetrahedral $Fe^{3+}$) $<$ $\delta$(octahedral $Fe^{3+}$) $\ll$ $\delta$(tetrahedral $Fe^{2+}$) $<$ $\delta$(octahedral $Fe^{2+}$) and are therefore diagnostic of the oxidation state and co-ordination. This generalization is also valid for sulphide spinels except that the shifts are uniformly smaller because of increased covalency. Where quadrupole splitting occurs, that for the $Fe^{3+}$ cation is usually quite small.

The internal magnetic field is the most important of the three hyperfine interactions in these systems. The magnetically ordered $Fe^{3+}$ cation has a saturation magnetic flux density of between 49

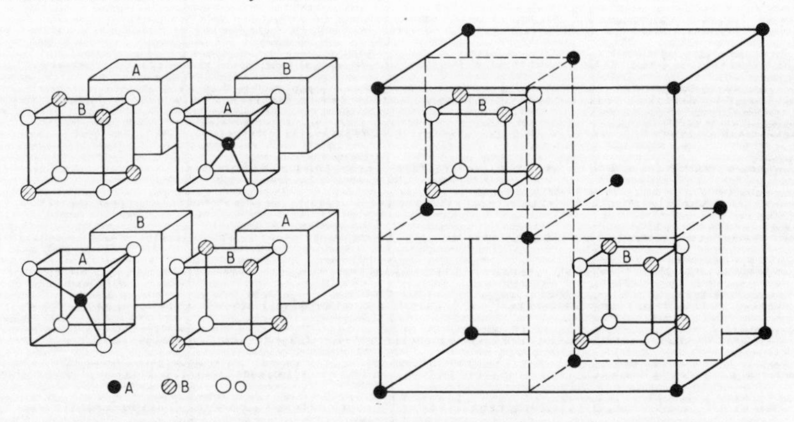


Fig. 7.1 The normal spinel structure, $A[B_2]O_4$, contains eight octants of alternating $AO_4$ and $B_4O_4$ units as shown on the left; the oxygens build up into a face-centred cubic lattice of 32 ions which co-ordinate A tetrahedrally and B octahedrally. The unit cell, $A_8[B_{16}]O_{32}$, is completed by an encompassing face-centred cube of A ions, as shown on the right in relation to two $B_4O_4$ cubes.

and 54 T (490–540 kG), essentially originating in the Fermi contact term $B_S$, and is usually much greater than any field from $Fe^{2+}$ which is subject to the orbital and dipolar contributions, $B_L$ and $B_D$. An $Fe^{3+}$ cation at a tetrahedral site will have a smaller magnetic flux density than at an octahedral site because of the greater covalency, but the difference is only of the order of 5% and may not be fully resolved.

$Fe^{3+}$ cations at the trigonally distorted B sites in a spinel show a quadrupole splitting in the paramagnetic state. For an *S*-state ion, this can only arise from the lattice contributions to the electric field gradient. A number of calculations of this lattice term have been made, but are often of dubious value. As an example, the normal spinels $Zn[Fe_2]O_4$ and $Cd[Fe_2]O_4$ show quadrupole splittings of 0.333 and 0.284 mm $s^{-1}$ respectively at 298 K [1]. The sign of $e^2qQ$ has been determined by applying an external magnetic field and in both cases is negative. However, a simple monopole calculation of $e^2qQ$, using point-charge summations based on the known positions of the ions in the crystal structure, gives completely the wrong sign. It transpires that in this instance the dipole polarizability of the oxygen anion gives the major contribution to the electric field gradient, which is of opposite sign to the formal ionic charge summation. The latter is therefore of no value in isolation. The Fe-O bond distance is shorter in the zinc spinel. This, together with the decrease of 0.018 mm $s^{-1}$ in the chemical isomer shift from the value in the cadmium spinel, indicates a higher degree of covalent bonding in $Zn[Fe_2]O_4$.

$Fe^{2+}$ cations at B sites show a large quadrupole splitting consistent with the trigonal symmetry (see Chapter 5). The $^5A_{1g}$ singlet level ($d_{z^2}$) lies lowest, and is the ground-state for the sixth *d*-electron. Thermal population of the $^5E_g$ excited level causes a temperature dependence of the quadrupole splitting. Data for $Ge[Fe_2]O_4$ between 77 and 1020 K are illustrated in Fig. 7.2, together with the theoretical fit [2] for a $^5A_{1g}$–$^5E_g$ separation of $\delta = 1145$ $cm^{-1}$ ($\equiv$13.7 kJ $mol^{-1}$; $\delta/k = 1650$ K). Distortions from regular octahedral symmetry in these oxides are usually much

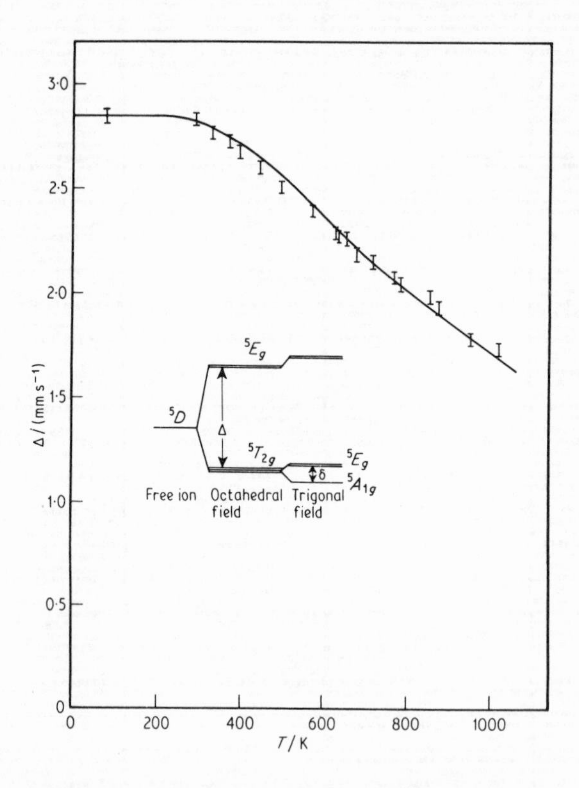


Fig. 7.2 The temperature dependence of the $^{57}Fe$ quadrupole splitting in $Ge[Fe_2]O_4$. The solid line corresponds to a $^5A_{1g}$-$^5E_g$ separation of $\delta = 1145$ $cm^{-1}$ ($\equiv$13.7 kJ $mol^{-1}$, $\delta/k = 1650$ K). ([2], Fig.1).

greater than in co-ordination complexes. Thus a particularly large splitting is also seen in the $^{99}Ru$ Mössbauer spectrum [3] of $Co^{2+}[Co^{III}Ru^{III}]O_4$. The low-spin $d^5$ $Ru^{3+}$ ion in this oxide has a similar level scheme to $Ge[Fe_2]O_4$, but with a filled $^2A_{1g}$ level and an electron 'hole' in the $^2E_g$ level.

$Fe^{2+}$ cations at A sites behave somewhat differently. $Fe[Cr_2]O_4$ is a cubic spinel phase at high temperatures, but transforms to a tetragonal symmetry below 135 K. The room temperature spectrum is a single line compatible with $Fe^{2+}$ ions at the regular tetrahedral A site [4]. As the temperature is lowered, a small quadrupole splitting appears at ~170 K, indicating a departure from cubic symmetry, and increases steadily until the Curie·temperature is reached at 90 K. This is illustrated in Fig. 7.3. A tetragonal distortion will lower the symmetry of the A site, but it is noteworthy that the process begins well *above* the nominal crystallographic transition temperature, in the vicinity of which the linewidths are particularly broad. Similar behaviour is found in $Fe[V_2]O_4$. The $Fe^{2+}$ ion is a Jahn–Teller ion, and should be unstable in a regular tetrahedral co-ordination. It is assumed that each $Fe^{2+}$ distorts its surroundings into a tetragonal symmetry, but that at high temperatures the direction of its tetragonal axis reorients dynamically among the three equivalent directions, thus maintaining the effective cubic structure. The reorientation frequency will be temperature-dependent. At temperatures well below the crystallographic phase transition the reorientation is slower than the observation time of the Mössbauer event so that the electric field gradient is observed. The broad lines at the transition region are directly due to this dynamic process.

Comparable behaviour is not found in $FeAl_2O_4$ because this spinel is partly inverse and does not therefore have strict cubic symmetry at the A site. It appears that two different situations can be defined. In a spinel where the B-site cations are all identical or are very similar, then the A-site symmetry remains cubic at high temperatures by dynamic motion and gives a static distortion on cooling. If the B-site cations are dissimilar, and this frequently means a difference in ionic charge, then there is a static distortion at the A site with a large electric field gradient. This applies to $FeAl_2O_4$ and also to the inverse spinel $Fe^{2+}[Fe^{2+}Ti^{4+}]O_4$. In both cases the static A-site distortion persists to beyond 800 K [5].

$FeV_2O_4$ and $FeCr_2O_4$ both become magnetic at temperatures

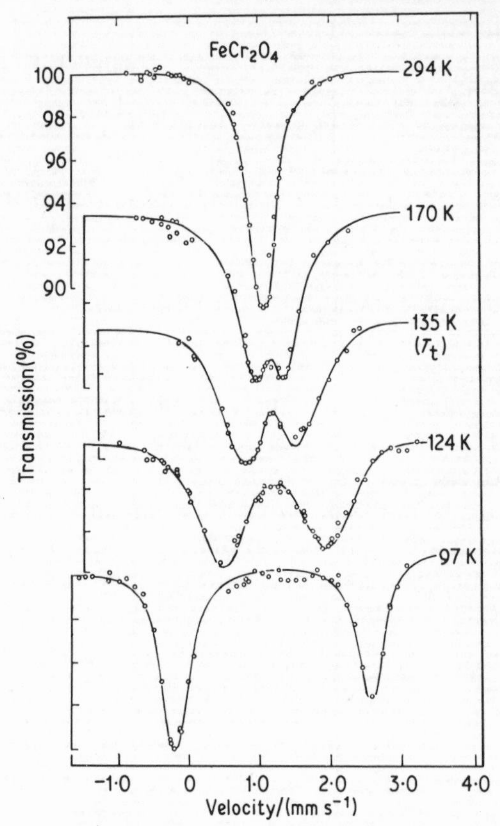


Fig. 7.3 Mössbauer spectra of $Fe[Cr_2]O_4$ showing the quadrupole splitting associated with the tetragonal distortion. ([4], Fig.3).

well below the cubic-tetragonal transition, but a more curious behaviour is found in for example $Fe[Cr_2]S_4$ which is cubic with an unsplit resonance line above its ordering temperature of 180 K. In the ferrimagnetic phase below 180 K there is a quadrupole splitting which appears to be magnetically induced [6]. This behaviour is not unexpected, but proves difficult to interpret theoretically.

A detailed discussion of magnetically ordered spinels will be deferred for the moment until Section 7.3 on exchange interactions.

## 7.2 Non-stoichiometric spinels

The introduction of non-stoichiometry causes many additional problems in the study of oxides. The cation site symmetry may no longer be unique, and the Mössbauer spectrum recorded is then a weighted summation which can no longer be analysed explicitly. In the spinel structure for example it is possible to tolerate a large number of vacant cation sites which we can represent as □. The classic example of this is in the spinel $\gamma$-$Fe_2O_3$ which, as there are 32 oxide ions in the unit cell, may be written as $Fe_{21\frac{1}{3}}\square_{2\frac{2}{3}}O_{32}$. The A and B sublattices in this material are magnetically coupled anti-parallel to each other in the ferrimagnetic phase. The A and B-site magnetic hyperfine flux densities can be distinguished by the application of an external field, $B_0$. The two internal fields align with $B_0$ (see Chapter 2) so that the observed flux densities are $B_A^{eff} = B_A + B_0$ and $B_B^{eff} = B_B - B_0$. The relative intensities of the A- and B-site patterns can be used to estimate the site occupancy ratio [7]. For a thin absorber the resonance intensity is approximately proportional to the concentration, and the recoilless fractions of the two sites are similar. Thus the site occupancy ratio can be deduced to be A/B = 0.62 ± 0.05. This result indicates the structure to be $(Fe)[Fe_{5/3}\square_{1/3}]O_4$ for which A/B = 0.6; that is, all the vacancies occur at the B sites. This procedure has been widely used to deduce site occupancies in other oxides, and further examples will be found later in the chapter. The flux densities of $B_A = 48.8$ T and $B_B = 49.9$ T at room temperature, together with the chemical isomer shifts relative to iron metal, $\delta_A = 0.27$ mm s$^{-1}$ and $\delta_B = 0.41$ mm s$^{-1}$, illustrate the generalizations made earlier on p. 160.

A second possible way of incorporating non-stoichiometry is by oxygen deficiency, although this is not a usual feature in spinels. However, a particularly clear example of this is found in the perovskite phases derived from the iron(IV) oxide $SrFeO_3$. The exact composition is very dependent on the experimental details of the preparation; for example, in $SrFeO_{2.86}$ the loss of oxygen is compensated by a partial reduction to $Fe^{3+}$. This results in a tetragonal distortion of the cubic perovskite lattice and the presence in the Mössbauer spectrum of contributions from both oxidation states [8]. Charge compensation in $SrFeO_{2.86}$ requires the formation of two $Fe^{3+}$ cations for each oxygen vacancy, and there is some evidence that these are closely related as a 'defect cluster' in the lattice. This may explain why the $Fe^{3+}$ ion shows a remarkably low magnetic flux density of only 42 T in the magnetically ordered phase at 4.2 K.

The conditions of preparation, and the subsequent thermal history of an oxide sample are very important when there are multiple site symmetries and/or a defect structure. For example in the spinel $CoFe_2O_4$, a slowly cooled sample has been prepared with only 7% inversion, whereas a quenched sample had 24% inversion [9]. Apart from the obvious effect on the relative intensities of the A and B sublattice spectra, there are also more subtle changes which are extremely important with regard to the magnetic properties. For this reason the effects of partial inversion and isomorphous replacement in general will be discussed more fully in connection with exchange interactions in Section 7.3.

It should be noted that the Mössbauer spectrum is not particularly sensitive to small concentrations of lattice defects. Even in the case of $\gamma$-$Fe_2O_3$ the presence of a large proportion of B-site vacancies is seen only as a difference in site occupancy, although in this example the $Fe^{3+}$ is expected to be particularly insensitive because it is an *S*-state ion. The primary effect of random fluctuations in the resonant site environment on the spectrum of the paramagnetic phase is usually a small degree of line broadening. For this reason the doping of oxides with impurity atoms of for example $^{57}Fe$ is not as successful a technique as one might have hoped.

One of the few systems where a systematic study of the effects of a defect structure has been made is in the binary oxide, FeO [10]. This oxide has the rock-salt face-centred cubic lattice, but is metastable at room temperature. Its preparation involves quenching from high temperatures and results in cation deficiency, so that the formula can be written as $Fe^{2+}_{1-3x}Fe^{3+}_{2x}\square_{x}O$. Samples between $Fe_{0.947}O$ and $Fe_{0.860}O$ all show an asymmetrical doublet spectrum at room temperature with a splitting of about 0.93 mm $s^{-1}$. The $Fe^{3+}$ ions appear to give a distinct contribution to the spectrum which is however not resolved. The $Fe^{2+}$ environment is distorted from the ideal cubic symmetry by the presence of neighbouring defects (hence the quadrupole splitting), but the random nature of the structure appears to result in a line broadening which cannot be simply interpreted. For this reason the magnetically ordered samples at low temperatures also show complex spectra.

Thermal annealing of $Fe_{1-x}O$ for short periods at temperatures above 400 K leads to partial decomposition into $Fe_3O_4$, but with the formation of a residual FeO phase which is very close to stoichiometry. The reaction can be represented as

$$(1-4z)Fe_{1-x}O \rightarrow (1-4x)Fe_{1-z}O + (x-z)Fe_3O_4$$

This disproportionation can be seen quite clearly in the Mössbauer spectrum (Fig. 7.4) because the quadrupole splitting and line broadening of the $Fe_{1-x}O$ phase decrease until only a single resonance line compatible with a cubic phase is seen superimposed

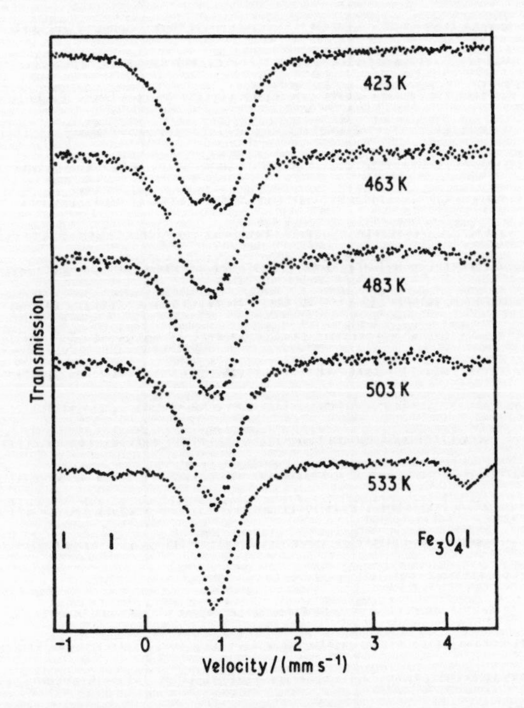


Fig. 7.4 The Mössbauer spectrum of $Fe_{0.940}O$ recorded during the precipitation of $Fe_3O_4$ by disproportionation. The sample was held at successive temperatures for about 2–3 hours. Note the gradual appearance of an intense line due to $Fe_3O_4$ on the right, and the narrowing of the $Fe_{1-x}O$ line as the phase tends towards stoichiometry. The line components due to $Fe_3O_4$ are indicated. ([10], Fig. 2).

upon the growing $Fe_3O_4$ resonance. Variations in the intensity of the various spectra with time enable the kinetics of the reactions to be followed. The spectra of $Fe_{1-x}O$ at 1074 K where the compound is thermodynamically stable comprise an unsplit broad line because of rapid diffusion of the vacancies (see Chapter 6).

## 7.3 Exchange interactions in spinels

The study of magnetic phases by Mössbauer spectroscopy has attracted considerable attention because it has permitted a much more detailed study of the mechanisms of spin-ordering and of the contributions of different sites to the sublattice magnetization. More recently, the effects of randomization of cations have been analysed in detail, and some of the properties of typical magnetic phases will now be described.

There are two aspects to consider, firstly the nature of the magnetic ordering and the configuration of the interacting spins, and secondly the magnitude of the interactions. Both may be determined from the Mössbauer spectrum. In the simple spin arrangements for ferromagnetic, ferrimagnetic, or antiferromagnetic ordering, the one or more sublattices show a collinear spin coupling. Thus a simple collinear antiferromagnet has two sublattices with antiparallel spin alignment (Néel ordering). In such an instance it is sufficient to determine the direction of spin alignment at any one site. For example if the site is distorted and there is a large electric field gradient at the nucleus which may be related to the geometry of the site, then the combined magnetic-quadrupole interaction may contain sufficient information to define the direction of the spin axis. An example of this in $FeF_2$ was given in Chapter 5. If on the other hand a single crystal is available, the relative intensities of the magnetic lines are angular-dependent and give the result more directly.

However, many oxide systems and particularly those which are nonstoichiometric do not have a collinear spin system, and a more detailed knowledge of spin orientation is a pre-requisite for the understanding of the magnetic properties in general. One of the more complicated ordering arrangements in spinels is the Yaffet–Kittel triangular ordering arrangement, in which there are effectively six sublattices in the antiferromagnetic structure. A further possibility is a canting of the spin away from the magnetic axis, which in a ferrimagnet may reduce the total magnetization (but not the moment at each site) by effectively introducing an antiferromagnetically coupled component in the perpendicular plane. Alternatively, spin canting in an antiferromagnet can introduce a small ferromagnetic contribution which becomes the major term in the susceptibility. Helical spin arrangements in which the spin is canted from the magnetic axis and successive spins along it precess in a spiral arrangement are common in antiferromagnets, an example of this being in the perovskite $SrFeO_3$ mentioned earlier.

The exchange interactions which lead to ordering usually take place via an intermediate oxygen ion and are therefore highly directional. The coupling of two magnetic ions depends more upon the M-O-M bond angle than upon the M-M distance. Thus in the spinel oxides the A-B site coupling far exceeds the A-A or B-B exchange. This results in some magnetically concentrated oxides such as $Zn[Fe_2]O_4$ and $Ge[Fe_2]O_4$ being paramagnetic except at extremely low temperatures (<20 K), while $MgFe_2O_4$ which is partly inverse and therefore has some A-B interactions is ferrimagnetic at room temperature. The value of the Néel or Curie temperature is in proportion to the strength of the exchange interactions, but more detailed information on the directional nature of the exchange is usually obtained by magnetic dilution. Thus the phase $(Mn_x^{2+}Zn_{1-x}^{2+})[Fe_2^{3+}]O_4$ will show a progressive increase in the number of A-B interactions as $x$ increases because $Mn^{2+}$ is a magnetic cation, whereas $Zn^{2+}$ is diamagnetic.

Some of the recent Mössbauer data for substituted spinels have shown conclusively that the magnetic field of an $Fe^{3+}$ cation is dependent on the nearest-neighbour cation environment, particularly when it occupies the B site. For example $NiFe_2O_4$, which is completely inverse and therefore has $Fe^{3+}$ at all six A sites about each B site, shows very sharp resonance lines, whereas $CoFe_2O_4$, $MnFe_2O_4$ and $MgFe_2O_4$ which are all partly inverse show a broadening of the hyperfine lines from the B sites due to the variations in the cation distribution at the nearest A sites [9]. A slowly cooled sample of $CoFe_2O_4$ with a cation distribution of $(Co_{0.07}Fe_{0.93})[Co_{0.93}Fe_{1.07}]O_4$ shows much less broadening than a quenched sample with a distribution of $(Co_{0.24}Fe_{0.76})[Co_{0.76}Fe_{1.24}]O_4$. A detailed analysis of the lineshapes from this sample has been made. The probability of finding $n$ $Co^{2+}$ ions on the six nearest-neighbour

A sites when the site occupancy is truly random is given by the binomial distribution

$$P(n) = \frac{6!}{n!(6-n)!}(1-c)^{6-n}c^n$$

where $c$ is the fractional occupation of the site in the bulk sample. For a value of $c = 0.24$, $P(0) = 0.19$, $P(1) = 0.37$, $P(2) = 0.29$ and $P(3) = 0.12$. The measured magnetic flux densities at room temperature were $B(0) = 51.5$, $B(1) = 49.9$, $B(2) = 47.5$ and $B(3) = 44.5$ T with a flux density at the A site of 49.9 T. The analysis of one of the component lines, that at extreme negative velocity, is shown in Fig. 7.5. A similar analysis has been made for $MnFe_2O_4$. These differences in flux density have been attributed to a supertransferred hyperfine interaction, that is a transfer of spin-imbalance from one metal ion to another via the oxygen anion [11]. The details are too involved to give here, but this phenomenon is reintroduced later in the chapter in a different context (see p. 179).

Very complex spectra are found for magnetic oxides in which

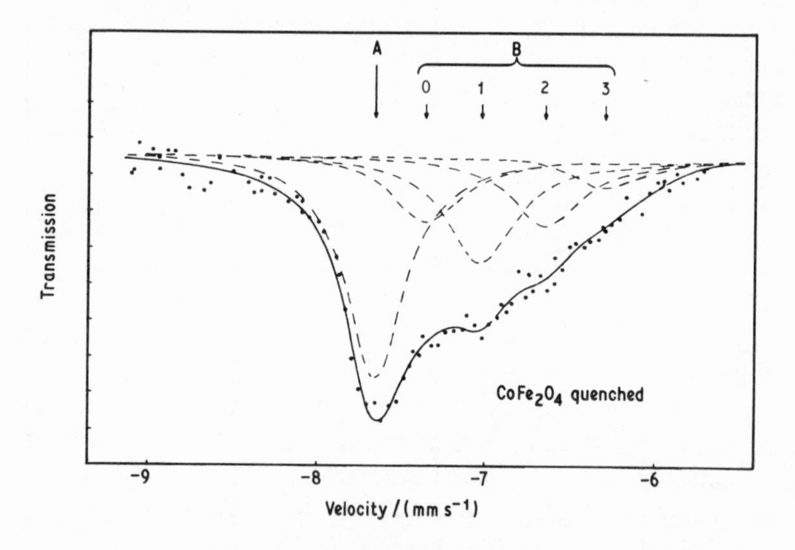


Fig. 7.5 One of the lines in the Mössbauer spectrum of $CoFe_2O_4$ (quenched sample) in an applied flux density of 1.7 T. The A site is relatively unaffected by the cation distribution, but contributions from iron atoms at B sites with 0, 1, 2 and 3 $Co^{2+}$ cations on the six nearest neighbour A sites are identified. ([9], Fig. 4).

the magnetic cations are diluted by a non-magnetic substituent. At low temperatures the spectrum of such an oxide containing $Fe^{3+}$ ions is usually a six-line spectrum, if somewhat broadened, but as the temperature increases there tends to be a progressive inward collapse, often with the appearance of an apparently paramagnetic central component before the ordering temperature is reached. This is illustrated in Fig. 7.6 for some data for $(Zn_{0.5}Fe_{0.5})[Co_{0.5}Fe_{1.5}]O_4$ where the most probable cation distribution is of three $Fe^{3+}$ and three $Zn^{2+}$ cations about each B site [12]. However, if the cation distribution is truly random, no less than 11%

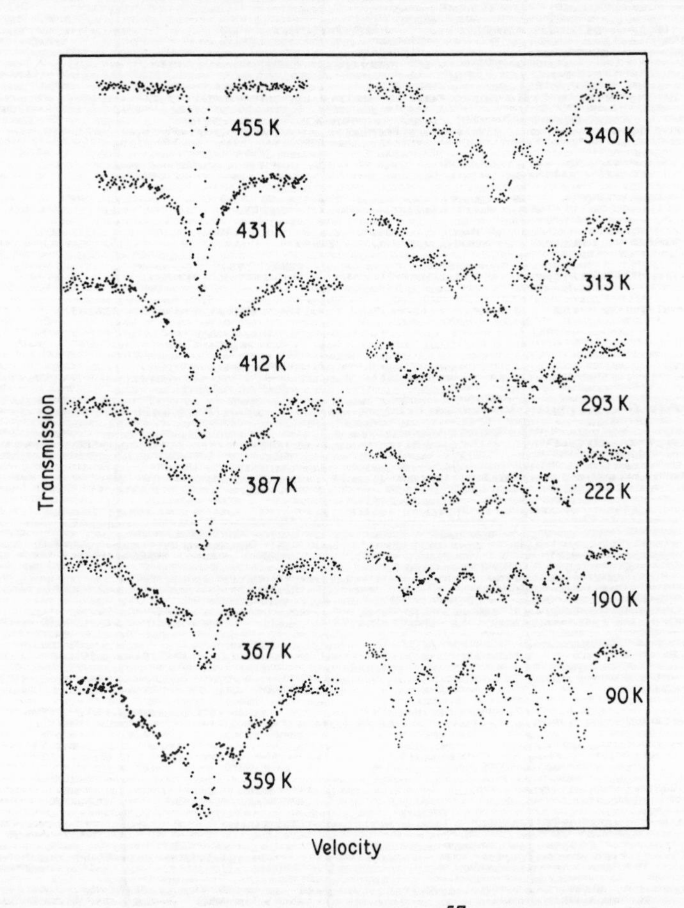


Fig. 7.6 The temperature dependence of the $^{57}Fe$ Mössbauer spectrum of $(Zn_{0.5}Fe_{0.5})[Co_{0.5}Fe_{1.5}]O_4$ showing the inward collapse of the hyperfine splitting with temperature increase and the appearance of a 'paramagnetic' line below the Néel temperature of 448 K. ([12], Figs. 1, 2).

of the B sites have only one or no iron neighbours, and if occupied by iron can be coupled only weakly to the lattice.

The spectra can be simulated by a stochastic model for spin fluctuations which gives excellent agreement with experiment over the whole temperature range. It therefore seems likely that those $Fe^{3+}$ cations which are only weakly coupled to the lattice are experiencing a fast relaxation well below the Curie temperature.

However, it is possible to explain at least some of the broadening without invoking relaxation. From molecular field theory it is possible to show that those $Fe^{3+}$ cations which have fewer neighbours will have a different temperature dependence of the hyperfine field [13]. In particular, the initial rate of decrease in the hyperfine field is greater and results in the observed inward broadening. This effect has a similar appearance to motional narrowing, but even for high magnetic dilution, as in the present example, it seems that the initial change in shape of the spectrum that accompanies the rise in temperature need not invoke a dynamic spin fluctuation. However, the molecular field theory approach appears to be inadequate in the region just below the ordering temperature.

The spinel phase $Co_{1-x}Zn_xFe_2O_4$ also serves as an example of the determination of a complex spin configuration [14]. The $Zn^{2+}$ preferentially enters A sites as in $Zn[Fe_2]O_4$, but the degree of inversion in ferrimagnetic $CoFe_2O_4$ is dependent on thermal treatment of the sample. $Fe^{3+}$ is an *S*-state ion, and this means that whatever the zero-field direction of the magnetic sublattice, it can be easily rotated by the application of an external field with which it aligns parallel or antiparallel. However, antiparallel Néel-coupled sublattices for example remain antiparallel (except in extremely large applied fields) so that the respective flux densities are increased or decreased by the value of the applied field. Even where the flux densities of different sublattices are very similar, they can often be distinguished by appropriate choice of the applied field strength (this method has already been described in connection with $\gamma$-$Fe_2O_3$). If spin-canting occurs, it cannot be easily destroyed, and although the resultant magnetic axis may align with the field, the individual spin axis will remain canted. This provides a means of distinguishing different cases. If an external field is applied to an $Fe^{3+}$ oxide to align the magnetic axis parallel to the direction of observation, the $\Delta m = 0$ components of the Mössbauer magnetic splitting should have zero intensity. This only remains true if each individual spin moment lies along this axis; any canting from the direction of observation will introduce a finite intensity into the $\Delta m = 0$ lines.

In the present example, the application of an external field of flux density 8 T to $CoFe_2O_4$ separates the A and B sublattice hyperfine lines, providing a means of determining the A and B site occupancy and of detecting spin canting. The spectra for $x = 0$ and $x = 0.6$ at 4.2 K are shown in Fig. 7.7. The variation in the intensities of the ouer lines can be used to determine the site occupancies which are $(Fe_{0.81}Co_{0.19})[Co_{0.81}Fe_{1.19}]O_4$ and

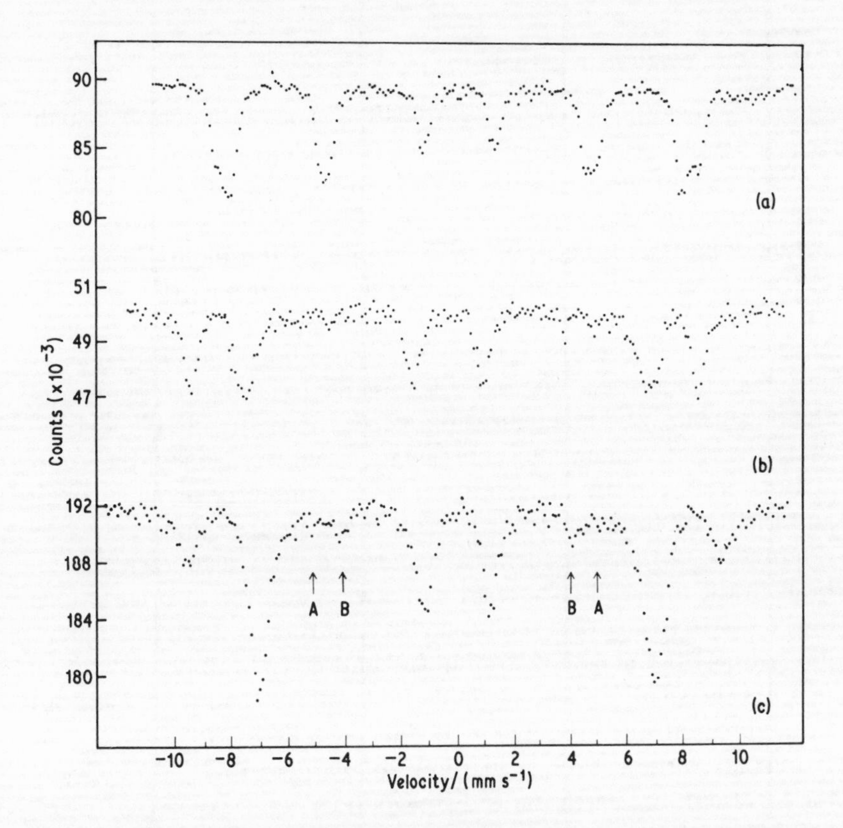


Fig. 7.7 $^{57}Fe$ Mössbauer spectra for $CoFe_2O_4$ at 4.2 K in (a) zero magnetic field (b) a flux density of 8 T applied along the direction of observation and (c) $Co_{0.4}Zn_{0.6}Fe_2O_4$ at 4.2 K in an externally applied flux density of 8 T. The variation in the intensities of the outer lines allows a determination of the site occupancy, and the weak intensity of the $\Delta m = 0$ lines in (b) and particularly in (c) indicates the presence of spin canting. ([14], Figs. 1, 3).

$(Zn_{0.6}Fe_{0.4})[Co_{0.4}Fe_{1.6}]O_4$. As can be seen the cobalt is quickly displaced from the A sites by zinc. The $\Delta m = 0$ lines are weakest in $CoFe_2O_4$, but increase rapidly in intensity above $x = 0.4$, and are very intense for $x = 0.8$ (not shown). Although $CoFe_2O_4$ approximates to a Néel ordered ferrimagnet, spin canting is present at both sites and increases with zinc substitution. Detailed study of the lineshapes for $x = 0.8$ provides evidence that individual B-site spins become increasingly canted as the number of $Zn^{2+}$ ions in the nearest neighbour A sites increases. Therefore in any given sample, the spins are canted by different angles with a statistical distribution.

## 7.4 Rare-earth iron garnets

Some good examples of the complex exchange interactions which can be studied are provided by the rare-earth iron garnets, $R_3Fe_5O_{12}$, and their substituted derivatives. The large unit cell contains 24 dodecahedral (c) sites occupied by $R^{3+}$ cations, 16 octahedral (a) sites occupied by $Fe^{3+}$, and 24 tetrahedral (d) sites also occupied by $Fe^{3+}$ cations; it is conventionally written as $\{R_3\}[Fe_2](Fe_3)O_{12}$. The structure is represented in Fig. 7.8 [15]. Mössbauer spectroscopy is particularly useful in complicated systems such as these because it is possible to study independently the occupancy and magnetizations of all three sites, and to determine the relative magnitudes of the different exchange interactions.

Both iron sites are distorted, the a-site symmetry being trigonal and the d-site tetragonal. As a result there are four differently oriented but otherwise identical a-sites and three d-site orientations in the unit cell. These are not always equivalent orientations in the magnetically ordered phase because there is then a unique direction of the magnetization in the unit cell. In polycrystalline samples the small $Fe^{3+}$ quadrupole interactions tend to average out as an apparent line broadening of the magnetic lines.

In some garnets, such as yttrium iron garnet (YIG), $Y_3Fe_5O_{12}$, the tervalent rare-earth ion is diamagnetic so that exchange forces concern solely the a and d sites, but in others such as the samarium, europium and dysprosium iron garnets, the rare-earth also has a sublattice magnetization, and the potential number of exchange interactions is much greater. The relative importance of these can be determined by selective magnetic dilution with a non-magnetic

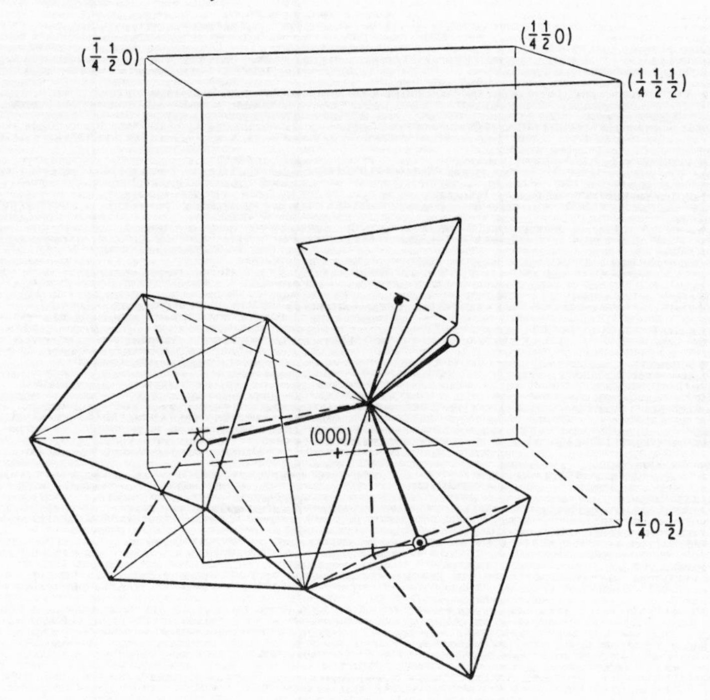


Fig. 7.8 A schematic representation of part of the unit cell of the $\{R_3\}[Fe_2](Fe_3)O_{12}$ lattice showing the inter-relation of the three cation site symmetries. (after [15]).

cation. Thus in the phase $\{Y_3\}[Fe_2](Fe_{3-x}Al_x)O_{12}$ the number of a-a interactions remains constant, whilst the numbers of a-d and d-d interactions decrease as the aluminium content increases.

As a general rule the major exchange coupling is between a and d sites, and gives a ferrimagnetic anti-parallel coupling of the two sublattices. The rare earth is then weakly coupled anti-parallel to the *resultant* moment of the iron sublattices, but the dominance of the a-d coupling results in the Curie temperature being close to 560 K regardless of whether or not the rare earth is magnetic. However, some of the finer details of the magnetic ordering in a stoichiometric iron garnet *are* dependent on the rare-earth. For example, in $Y_3Fe_5O_{12}$ the magnetization lies along the [111] direction so that there are two non-equivalent octahedral a-site orientations with all tetrahedral d sites equivalent. The corresponding $^{57}Fe$ spectrum at 300 K shows magnetic flux densities of 49.0

and 48.4 T at the a sites and 39.3 T at the d site with the relative intensities of 3:1:6 as predicted [16]. As with the spinels, the flux density at the tetrahedral site is much smaller than those of the octahedral sites. By contrast, in $Sm_3Fe_5O_{12}$ the magnetization is along the [110] direction and there are now two non-equivalent d-site orientations. The flux densities at 295 K in this case are 49.5 T at the a sites and 40.8 and 40.6 T at the d sites with relative intensities of 4:2:4. It is possible to distinguish the two d sites because they show different quadrupole splittings of $e^2qQ/2 = -0.20$ and +0.10 mm $s^{-1}$ respectively which perturb and separate the magnetic lines.

The importance of the a-d interaction can quite easily be seen from a comparison of Curie temperatures. For example $Y_3Fe_5O_{12}$ ($T_C$ = 550 K) and $Gd_3Fe_5O_{12}$ (560 K) are typical of garnets with a strong a-d exchange, whereas compounds with mainly a-a interactions such as $\{Ca_3\}[Fe_2](Si_3)O_{12}$ and $\{Ca_3\}[Fe_2](Gd_3)O_{12}$, or d-d interactions as in $\{YCa_2\}[Sn_2](Fe_3)O_{12}$ and $\{NaCa_2\}[Sb_2](Fe_3)O_{12}$ or c-c interactions as in $\{Fe_3^{2+}\}[Al_2](Si_3)O_{12}$ are all paramagnetic above 80 K [17].

Each a site has 6 exchange linkages via oxygen to d sites, and each d site has 4 linkages to a sites. Any given magnetic ion must have two or more linkages with magnetic cations in the other sublattice to maintain continuity in the ferrimagnetic ordering. Thus in the phase $\{Y_3\}[Fe_2](Fe_{3-x}Al_x)O_{12}$, when $x = 1.7$ 18% of the a-site $Fe^{3+}$ ions have either none or one exchange linkage.

As we have already seen, weakly coupled atoms show a large decrease in the magnetization and a consequent decrease in the internal magnetic flux density. This effect will be particularly important at temperatures close to the Curie temperature. For example in the phase $\{Y_{3-x}Ca_x\}[Fe_2](Fe_{3-x}Si_x)O_{12}$ there is a substantial inward collapse of the spectrum above $x = 1.5$ at room temperature (see Fig. 7.9) as the magnetic interactions weaken [18]. Note the reversal in the intensities of the outer pairs of lines between $x = 0.15$ and $x = 1.2$. The extreme outer lines are due to a sites and become more prominent as the contribution from the d sites is reduced. This provides a means of determining which site is substituted.

Although the ground state of $Eu^{3+}$ is nominally the non-magnetic $^7F_0$ level, crystal-field and exchange interactions mix in the $^7F_J$ excited states so that $Eu_3Fe_5O_{12}$ shows ordering at both

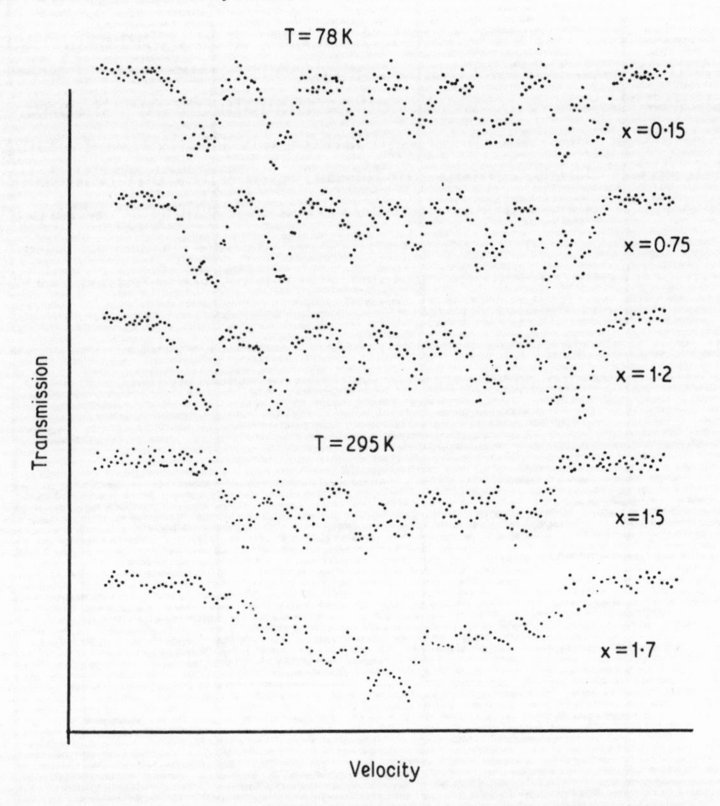


Fig. 7.9 Spectra of the garnets $Y_{3-x}Ca_xFe_{5-x}Si_xO_{12}$ showing how the magnetic hyperfine splitting is affected by the magnetic dilution for values of $x$ greater than 1.5. ([18], Fig. 1).

the europium and iron sites. All the c-sites are not in fact equivalent because the iron sublattice magnetization is in the [111] direction, and there are two $^{151}Eu$ hyperfine flux densities of 56.1 and 62.3 T at 4.2 K [19]. Substitution of gallium for iron takes place preferentially at the d sites to give $\{Eu_3\}[Fe_2](Ga_xFe_{3-x})O_{12}$. The exact site occupancy can again be determined from the $^{57}Fe$ spectra, and in fact 85% of the gallium is in d sites for $x = 0.66$ and 73% for $x = 3.03$ [20]. It is interesting to note that the a- and d-site magnetizations are opposed to each other, and the net magnetization vanishes at $x \sim 1.4$ because of 'magnetic compensation'.

The $^{151}Eu$ resonance at 4.2 K is altered dramatically with increase in $x$, and as can be seen in Fig. 7.10, the magnetic splitting collapses completely. The europium field is generated by an exchange interaction with the two nearest-neighbour Fe atoms at d sites, which will therefore be weakened by the gallium substitution. The observed spectra can be simulated successfully using a simple model. Each europium ion can have one of three environments: (a) two $Fe^{3+}$ neighbours at d sites (statistical weight $y^2$ where $y$ is the concentration of iron on d sites determined from the $^{57}Fe$ spectra); (b) one $Fe^{3+}$ and one $Ga^{3+}$ (weight $2y[1-y]$); (c) two $Ga^{3+}$ neighbours (weight $[1-y]^2$). The exchange mechanisms produce a flux density at the $Eu^{3+}$ of $B_c$, $0.52B_c$ and 0 respectively. The solid curves in the figure were calculated using essentially

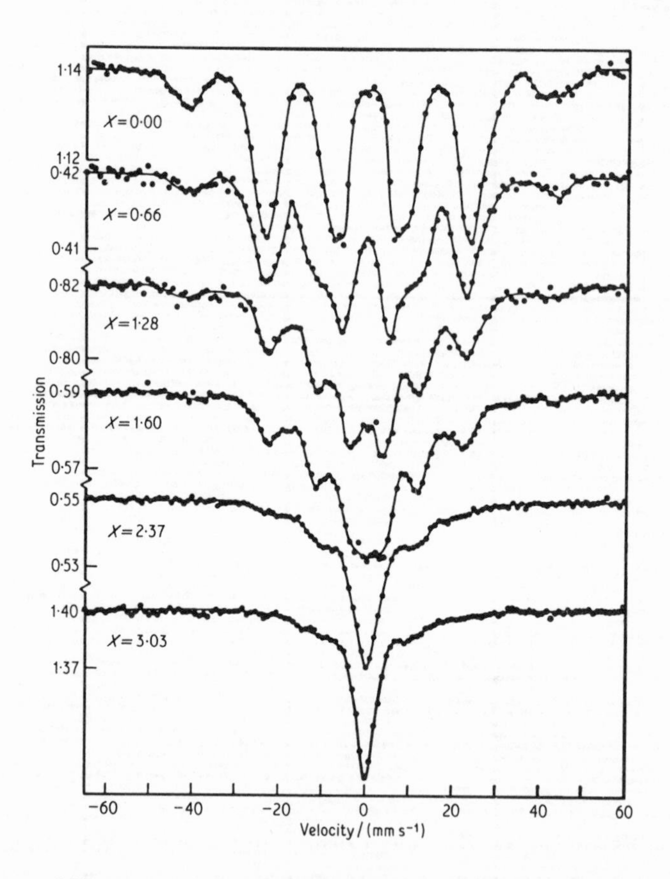


Fig. 7.10 $^{151}Eu$ spectra of gallium-substituted europium iron garnets of the type $Eu_3Fe_{5-x}Ga_xO_{12}$ showing the reduction in the exchange interaction at the $Eu^{3+}$ ion with increasing gallium substitution. ([20], Fig. 3).

this simple model, but including an additional contribution to the exchange field of 12% from the four third-nearest neighbours on d sites.

Scandium substitution takes place exclusively at the a sites, as in $\{Eu_3\}[Sc_xFe_{2-x}](Fe_3)O_{12}$, and one would expect that the Eu-Fe exchange via d sites would be unaffected. However, partial collapse of the $^{151}Eu$ magnetic spectrum is also seen here [21]. It appears that in this case the iron d-site spins are canted from the [111] direction by an angle determined by the number of a-d site exchange linkages, and as a result the c-d interactions show a statistical variation. Complex spin-canting can also be identified in $\{Eu_xSm_{1-x}\}[Fe_2](Fe_3)O_{12}$ where the magnetization is along the [111] direction for $x = 1$ and along the [110] direction for $x = 0$ [22].

It will now be clear that the effects of substitution on the magnetic exchange interactions can be very subtle and complex. Where multiple sublattices are involved, Mössbauer spectroscopy is uniquely successful in elucidating the nature of the operative mechanisms.

## 7.5 Transferred hyperfine interactions

Although a diamagnetic cation or anion has no resultant spin and therefore cannot order, it may nevertheless be influenced by the close proximity of another cation with a magnetically ordered spin. The electrons on the diamagnetic atom then become polarized by exchange interactions so that there is a small resultant imbalance in the spin density at the nucleus and consequently a magnetic hyperfine field. These transferred hyperfine interactions can be quite large, and in $^{119}Sn$ for example, magnetic flux densities of over 20 T have been found in some oxide systems.

An interesting example is given by the garnets $\{Y_{3-x}Ca_x\}[Fe_{2-x}Sn_x](Fe_3)O_{12}$. The $Sn^{4+}$ diamagnetic cations at the a site are all surrounded by six $Fe^{3+}$ cations at d sites. It was made clear in the preceding section that the a-d exchange between iron atoms is not appreciably affected by the initial magnetic dilution, and in this case there is no effect until $x > 0.9$. The $^{119}Sn$ spectrum shows a large magnetic flux density (Fig. 7.11) which, being produced by six $Fe^{3+}$ neighbours is insensitive to the degree of tin substitution below $x = 0.9$ [23, 24]. However, the breakdown of long-range

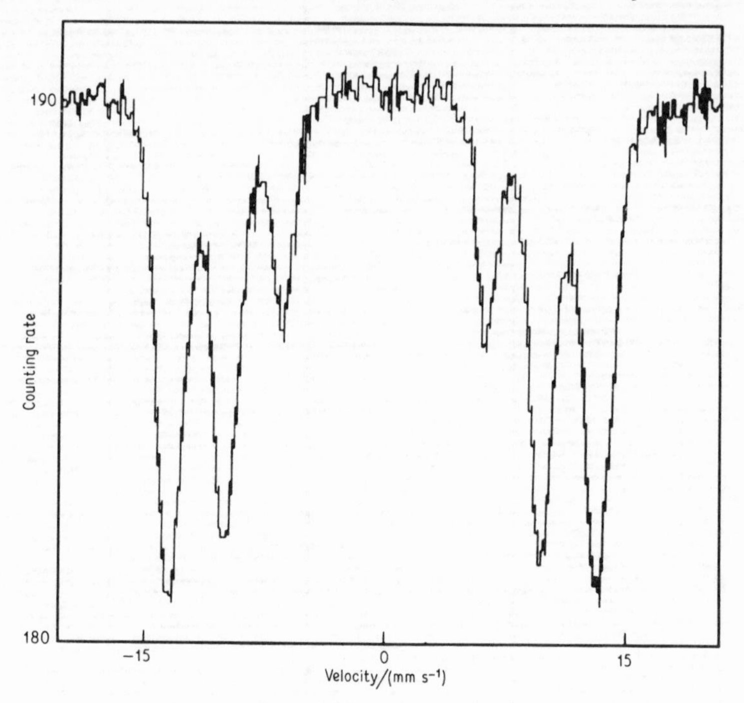


Fig. 7.11 The $^{119}Sn$ spectrum of $Ca_{0.25}Y_{2.75}Sn_{0.25}Fe_{4.75}O_{12}$ at 77 K showing a magnetic hyperfine splitting of over 20 T from a transferred hyperfine interaction. ([23], Fig. 1).

order between $x = 0.9$ and 1.2 results in a rapid collapse of the flux density at both the $^{57}Fe$ and $^{119}Sn$ nuclei.

The cation $Sb^{5+}$ is isoelectronic with $Sn^{4+}$, and transferred hyperfine interactions are also known for antimony in the spinels $Ni_{1+2x}Fe_{2-3x}Sb_xO_4$ [25]. The antimony substitutes at the octahedral B sites, but here the average number of $Ni^{2+}$ neighbours at A sites alters with increasing substitution. The result is a more complex spectrum in which each possible combination of A-site neighbours produces a different hyperfine flux density at the antimony.

Transferred hyperfine interactions can also occur at the anion, and an example of this can be found in the ferromagnetic spinel $CuCr_2Te_4$ where the $^{125}Te$ resonance reveals a flux density of 14.8 T at 80 K [26].

## References

[1] Evans, B. J., Hafner, S. S. and Weber, H. P. (1971) *J. Chem. Phys.*, **55**, 5282.
[2] Eibschutz, M., Ganiel, U. and Shtrikman, S. (1966) *Phys. Rev.*, **151**, 245.
[3] Gibb, T. C., Greatrex, R., Greenwood, N. N., Puxley, D. C. and Snowdon, K. (1973) *Chem. Phys. Letters*, **20**, 130.
[4] Tanaka, M., Tokoro, T. and Aiyama, Y. (1966) *J. Phys. Soc. Japan*, **21**, 262.
[5] Ono, K., Chandler, L. and Ito, A. (1968) *J. Phys. Soc. Japan*, **25**, 175.
[6] Eibschutz, M., Hermon, E. and Shtrikman, S. (1967) *J. Phys. Chem. Solids*, **28**, 1633.
[7] Armstrong, R. J., Morrish, A. H. and Sawatzky, G. A. (1966) *Phys. Letters*, **23**, 414.
[8] Gallagher, P. K., MacChesney, J. B. and Buchanan, D. N. E. (1964) *J. Chem. Phys.*, **41**, 2429.
[9] Sawatzky, G. A., van der Woude, F. and Morrish, A. H. (1969) *Phys. Rev.*, **187**, 747.
[10] Greenwood, N. N. and Howe, A. T. (1972) *J. Chem. Soc. (A)*, pp. 110, 116 and 122.
[11] van der Woude, F. and Sawatzky, G. A. (1971) *Phys. Rev.*, **4**, 3159.
[12] Iyengar, P. K. and Bhargava, S. C. (1971) *Phys. Stat. Sol. B*, **46**, 117.
[13] Coey, J. M. D. and Sawatzky, G. A. (1971) *Phys. Stat. Sol. B*, **44**, 673.
[14] Pettit, G. A. and Forester, O. W. (1971) *Phys. Rev. B*, **4**, 3912.
[15] Gilleo, M. A. and Geller, S. (1958) *Phys. Rev.*, **110**, 73.
[16] van Loef, J. J. (1968) *J. Appl. Phys.*, **39**, 1258.
[17] Lyubutin, I. S. and Lyubutina, L. G. (1971) *Soviet Physics-Crystallography*, **15**, 708.
[18] Lyubutin, I. S., Belyaev, L. M., Vishnyakov, Y. S., Dmitrieva, T. V., Dodokin, A. P., Dubossarskaya, V. P. and Shylakhina, L. P. (1970) *Soviet Physics-JETP*, **31**, 647.
[19] Stachel, M., Hufner, S., Crecelius, G. and Quitmann, D. (1969) *Phys. Rev.*, **186**, 355.
[20] Nowik, I. and Ofer, S. (1967) *Phys. Rev.*, **153**, 409.
[21] Bauminger, E. R., Nowik, I. and Ofer, S. (1967) *Phys. Letters*, **29A**, 328.
[22] Atzmony, U., Bauminger, E. R., Mustachi, A., Nowik, I., Ofer, S. and Tassa, M. (1969) *Phys. Rev.*, **179**, 514.
[23] Goldanskii, V. I., Trukhtanov, V. A., Devisheva, M. N. and Belov, V. F. (1965) *ZETF Letters*, **1**, 19.
[24] Belov, K. P. and Lyubutin, I. S. (1966) *Soviet Physics-JETP*, **22**, 518.
[25] Ruby, S. L., Evans, B. J. and Hafner, S. S. (1968) *Solid State Comm.*, **6**, 277.
[26] Ullrich, J. F. and Vincent, D. H. (1967) *Phys. Letters*, **25A**, 731.

CHAPTER EIGHT

# *Alloys and Intermetallic Compounds*

One of the major applications of Mössbauer spectroscopy is in the field of metal physics. Many of the physical properties of a metal such as the electrical conductivity and magnetic susceptibility are macroscopic; i.e. they derive from the bulk solid and in particular from its collective-electron band structure. The Mössbauer technique differs in that it records individual atoms in the metal. This enables a detailed study of the near-neighbour interactions with the resonant nucleus, and of the effects of changing composition on the electronic and magnetic interactions of particular atoms within the alloy. Many of the principles enunciated in earlier chapters are still applicable, but some new phenomena are found and the more important of these will now be discussed. Some of the analytical applications in applied metallurgy are indicated in Chapter 9.

Most interest is centred on magnetically ordered alloys, and in this context it is important to consider the origin of the internal magnetic field at the resonant nucleus. If the Mössbauer isotope is a transition metal such as iron, then the band-structure of the alloy may be such that there is still an effective localized *d*-electron magnetic moment on the atom. It is no longer possible to speak in terms of an integral number of *d*-electrons, as in Chapter 5, but this unpaired *d*-band spin-density can induce an imbalance in the *s*-electron spin-density at the nucleus, and hence a magnetic flux density equivalent to the Fermi interaction described on p. 110. This effect is known as 'core-polarization'. In addition, there may be a contribution from unpaired conduction-electron spin-density which usually has considerable *s*-character and therefore also contributes significantly to the observed magnetic flux density. This is referred to as the 'conduction-electron polarization'. The relative importance of these two effects and the way in which they are influenced by change in composition of the alloy can often be determined by Mössbauer spectroscopy. The temperature dependence of the magnetic flux density below the ordering temperature is essentially similar to the behaviour found in ionic compounds (see p. 112) and will not be discussed further.

It is convenient to consider any metallic phase as belonging to one of two classifications. A disordered alloy is one in which two or more elements occupy the same crystallographic sites with a random probability. Such a phase frequently has a wide range of composition. An intermetallic compound differs in that each element shows strong preference for a particular lattice site. The result is a regular structure with a composition close to a simple stoichiometry and a very restricted range of composition. However, there are intermediate examples where an alloy shows a variable amount of atomic order/disorder according to its previous thermal and mechanical history.

## 8.1 Disordered alloys

The chemical isomer shift and the magnetic flux density at the nucleus in an alloy are intimately related to the electron bandstructure. A systematic study of the variation of these parameters with change in composition, can therefore be expected to yield significant information about the structure of the alloy.

A good example of the use of the chemical isomer shift in nonmagnetic alloys is provided by the solid-solution of palladium with 0–16.5 at.% of tin. This alloy has the face-centred cubic structure [1]. The paramagnetic susceptibility of the alloy decreases rapidly with increasing tin content from the value in pure palladium, and approaches zero at the limit of solubility. Surprisingly, the $^{119}Sn$ chemical isomer shift is invariant over the whole range of composition ($-1.0$ mm s$^{-1}$ with respect to $\beta$-tin metal). This shift is also characteristically different from that in any of the tin-rich intermetallic compounds $Pd_3Sn$, $Pd_2Sn$, PdSn, $PdSn_2$ and $PdSn_4$, thereby indicating that the electronic configuration of the tin in the solid-solution phase is indeed independent of composition. The large negative value of the shift denotes a significant decrease in the 5*s*-electron density at the tin compared to the pure metal. The paramagnetic susceptibility of palladium can be attributed to a spin-moment from partial filling of the 4*d*-band. Both the magnetic and Mössbauer data for the alloy can be explained by a simple rigid band model in which the tin valence electrons are donated to the unfilled 4*d*-band of the palladium. The latter apparently becomes filled at the limit of solid solubility. The effectively complete loss of electrons from the 5*s*-band of the tin results in the large negative chemical isomer shift.

The itinerant nature and the high *s*-character of the binding conduction-electrons tend to minimize the effects of variations in the near-neighbour environment on the Mössbauer spectrum. Consequently, any influences of such variations on the quadrupole splitting or chemical isomer shift are usually obscured by the width of the Mössbauer line, and can be tacitly ignored. However, this is not always the case with magnetic interactions.

In a pure metal phase such as ferromagnetic $\alpha$-iron (body-centred cubic lattice) there is only one type of iron site. The $^{57}$Fe Mössbauer spectrum is therefore a single 6-line magnetic hyperfine pattern, the magnetic flux density of 33.0 T at 294 K originating largely from core-polarization. Consider in the first instance a dilute substitution of an impurity metal into the $\alpha$-iron structure. Each impurity atom may be expected to have a significant effect on the charge distribution and spin-density at the neighbouring iron sites. The effect on the Mössbauer spectrum will be a change in the observed magnetic flux density at all the iron sites affected which is dependent on the nature of the impurity atom and on its distance from the resonant atom. As a result, the observed $^{57}$Fe spectrum is a statistical summation of all the different contributions, and the single hyperfine pattern is replaced by a broadened envelope with only a poorly resolved fine structure.

Four examples of the spectra of alloys of $\alpha$-iron containing a small proportion of a second metal are shown in Fig. 8.1 [2]. The complicated line-shapes result directly from the multiple environments. The major experimental problem lies in the correct analysis of these line-shapes. A systematic survey [2] of the transition metals alloyed with iron has shown that Ti, V, Cr, Mn, Mo, Ru, W,

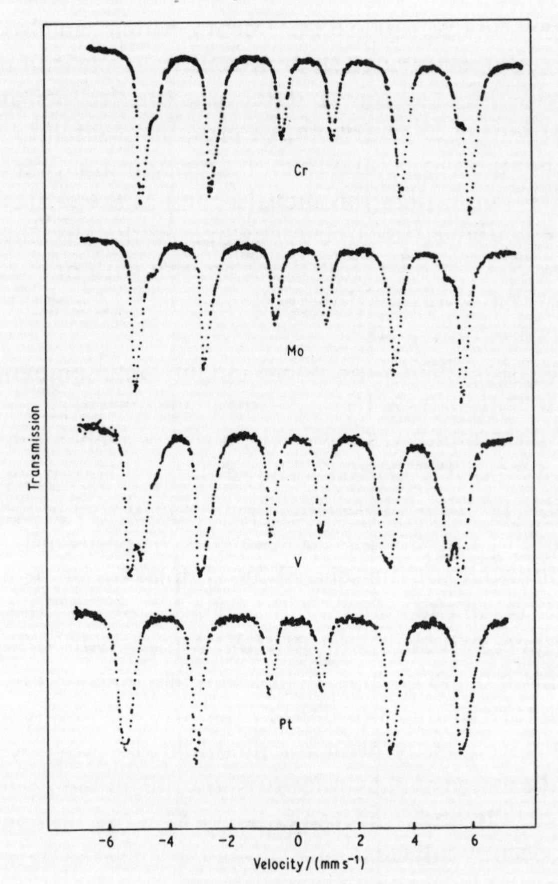


Fig. 8.1 The $^{57}$Fe spectra at room temperature of alloys derived from α-iron by substitution of (top to bottom) 2.2 at.% Cr, 1.0 % Mo, 5.0 % V, 3.0 % Pt. ([2], Fig. 1).

Re and Os impurities (i.e. those elements to the left of or below iron in the periodic table) give a substantial reduction in the effective hyperfine flux density with observable satellite lines in the spectrum. On the other hand, Co, Ni, Rh, Pd, Ir and Pt (i.e. elements to the right of iron) give only slight broadening but with an increased magnetic flux density. This is probably indicative of the relative effects on the iron 3*d*-band.

More detailed analysis is best achieved in conjunction with complimentary measurements by spin-echo and continuous-wave nuclear magnetic resonance experiments, which can also be used to study variations in magnetic flux-density at the iron nucleus.

It is usually only possible to determine the flux density for iron atoms contributing to the two most prominent satellite resonances (generally assumed to be due to an impurity in the first or second near-neighbour co-ordination shells; a total of 8 + 6 sites in the body-centred cubic lattice). For example in the Mo alloy, the field is reduced by 3.87 ± 0.05 T for an iron with one molybdenum atom in the first co-ordination shell, and by 3.16 ± 0.07 T for the second co-ordination shell.

A recently described experiment circumvents some of the problems in an ingenious way [3]. A single crystal foil of Fe containing 4.8 at.% of chromium was magnetized successively along the [111] and [001] axes of the crystal. The resulting differences between the two corresponding Mössbauer spectra lead to a more positive identification of the satellite components. It is suggested that it is the 1st and 5th (not the second) co-ordination shells which show large reductions in the magnetic flux density of –3.30 T and –2.37 T respectively, the 2nd, 3rd and 4th co-ordination spheres producing little or no change. These conclusions lead to an unexpected picture of the spin-density disturbance about the chromium. It has frequently been assumed that the effect will either decrease with increasing distance or show a damped oscillatory behaviour. In the body-centred cubic lattice, the first and fifth neighbours occupy adjacent sites along the diagonal axes of the cubic cell, e.g. (111) and (222). It would therefore appear that the spin-density disturbance is large along the diagonal axes (1st and 5th neighbours), but close to zero along the cube axes (2nd neighbour) and intermediate directions (3rd and 4th neighbours). It remains to be seen whether this directional model is applicable to any of the other dilute impurities in $\alpha$-iron.

Alloys with a higher concentration of the second element feature a significant probability that any iron atom has two or more impurity atoms in its first co-ordination shell, resulting in a large increase in the number of site environments. As just described, the gross effect of one nearest neighbour is to produce a shoulder (albeit of complex shape) on the main resonance peaks, corresponding to an average change in magnetic flux density of about 2 T. Those atoms with two, three, four etc. substituted neighbouring sites generate additional shoulders, the average change in flux density being approximately proportional to the number of neighbours. This behaviour is found for example in the disordered Fe-Si alloys (<10 at.% Si) where components corresponding to 0, 1 and 2 near-neighbours may be discerned [4].

Near-neighbour effects are not confined to the $^{57}$Fe resonance, although much less data are available for other isotopes. An interesting example is given by the nickel-palladium alloy system [5]. The magnetic flux density at the $^{61}$Ni nucleus in ferromagnetic nickel metal at 4.2 K is –7.6 T. The effective electronic moment per nickel atom as determined by magnetic measurements is 0.70 $\mu_B$. The Ni-Pd alloys are ferromagnetic over the whole of the concentration range 0–98 at.% Pd, the effective moment per nickel atom, rising to about 3$\mu_B$ in the Pd-rich alloys because of an induced magnetic effect on the palladium host.

The $^{61}$Ni Mössbauer spectra show considerable broadening of the hyperfine pattern in the intermediate concentrations, due to variations in the flux density caused by near-neighbour effects. Unfortunately the resolution of the fine-structure is inadequate for a detailed analysis. However, it is significant to note that the 'average' flux density changes in a regular manner, and reverses sign at about 53 at.% Pd. The data are illustrated in Fig. 8.2. The normal contributions to the magnetic flux density from core-polarization by the nickel 3*d*-moment and from conduction-electron polarization are both expected to be negative in sign in accord with measurements on other Ni-based alloys. However, the observed behaviour in the Ni-Pd alloys can only be explained if a third *positive* term is invoked which arises from a polarization of the nearest neighbour Pd atoms by the nickel magnetic moment via the conduction electrons. This contribution to the 'average' flux density observed is directly proportional to the 'average' number of Pd atoms in the first co-ordination shell of the nickel. The solid curve in Fig. 8.2 represents a theoretical model on this basis, and the good agreement with experiment confirms the idea that the bulk magnetic properties of the Pd-rich alloys are determined not by the d-band of the palladium host but by clusters of polarized Pd atoms about each Ni atom.

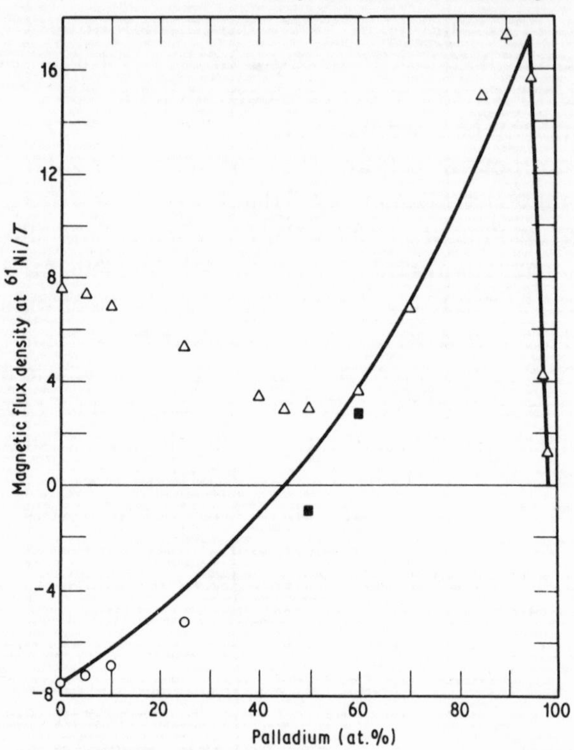


Fig. 8.2 The variation of the magnetic flux density at $^{61}$Ni at 4.2 K as a function of concentration in the Ni-Pd alloy system. The triangles represent the experimental average value of $|B|$. The circles are values of $B$ where the sign was observed to be negative. The squares represent estimated values from a distribution of fields. The solid line is a proposed theoretical model. ([5], Fig. 6).

## 8.2 Intermetallic compounds

A binary intermetallic compound usually has a restricted range of composition close to a simple stoichiometry such as $AB_4$, $AB_3$, $AB_2$, $A_2B_3$, AB etc., and has a correspondingly simple crystal structure in which the two elements occupy distinct crystallographic sites. In a sense these compounds are intermediate between the ionic lattice with localized electrons and the true random alloy. The uniformity of the atomic environments ensures that the Mössbauer spectrum no longer shows the line broadening found in a solid-solution alloy. Consequently, any small quadrupole interaction usually remains visible and can be related directly to the local site symmetry.

There has been considerable interest in the study of magnetically ordered intermetallic compounds, and it is not uncommon to find subtle complexities in the magnetic behaviour. A good example is given by the cubic Laves phases, $MFe_2$, where M is a rare-earth atom. The iron atoms are all crystallographically equivalent and occupy corner-sharing tetrahedral networks with a three-fold symmetry about the [111] directions. As a result, the electric field gradient at the iron is axially symmetric with $V_{zz}$ along this axis. The direction of the magnetic axis is determined by the rare-earth element. For example $DyFe_2$ and $HoFe_2$ give a single $^{57}Fe$ magnetic hyperfine pattern ($HoFe_2$ illustrated in Fig. 8.3) showing that the iron sites are all magnetically equivalent [6, 7]. This can only be true if the axis of magnetization is along [001] so that $V_{zz}$ has the same relative orientation at each site with respect to the magnetic flux density, *B*. In other phases such as $ZrFe_2$, $YFe_2$, $TbFe_2$, $ErFe_2$ and $TmFe_2$, the Mössbauer spectrum shows two very distinct magnetic patterns in the area ratios of 3:1 ($ErFe_2$ illustrated in Fig. 8.3). In these compounds the magnetic axis is along [111] so that $V_{zz}$ is parallel to $B$ for $\frac{1}{4}$ of the iron atoms and at 108° to $B$ for $\frac{3}{4}$ of the atoms.

A compound containing a magnetic element for which no Mössbauer resonance is available may often be studied indirectly via the transferred hyperfine interactions at a non-magnetic nucleus. For example some of the compounds of gold with the 3*d*-metals show ordering of the 3*d*-magnetic moments, which results in transferred magnetic hyperfine interactions in the $^{197}Au$ resonance. A particularly interesting series of examples are provided by some of the Au-Mn phases.

The $^{197}Au$ Mössbauer resonance in a solid-solution alloy of gold with 5 at.% manganese shows two superimposed hyperfine patterns corresponding to none or one manganese nearest neighbour [8]. The magnetic flux densities are 12.7 and 34.4 T respectively. The gold has no intrinsic moment, but the core and 6*s*-conduction electrons are spin-polarized by the neighbouring manganese moments. The magnetic flux density at the *isolated* gold atom is dependent on manganese concentration, and rises from 12.7 T at 5 at.% Mn to 16.5 T at 10 at.% Mn. The most likely origin of this field would therefore seem to be a *long-range* spin-polarization

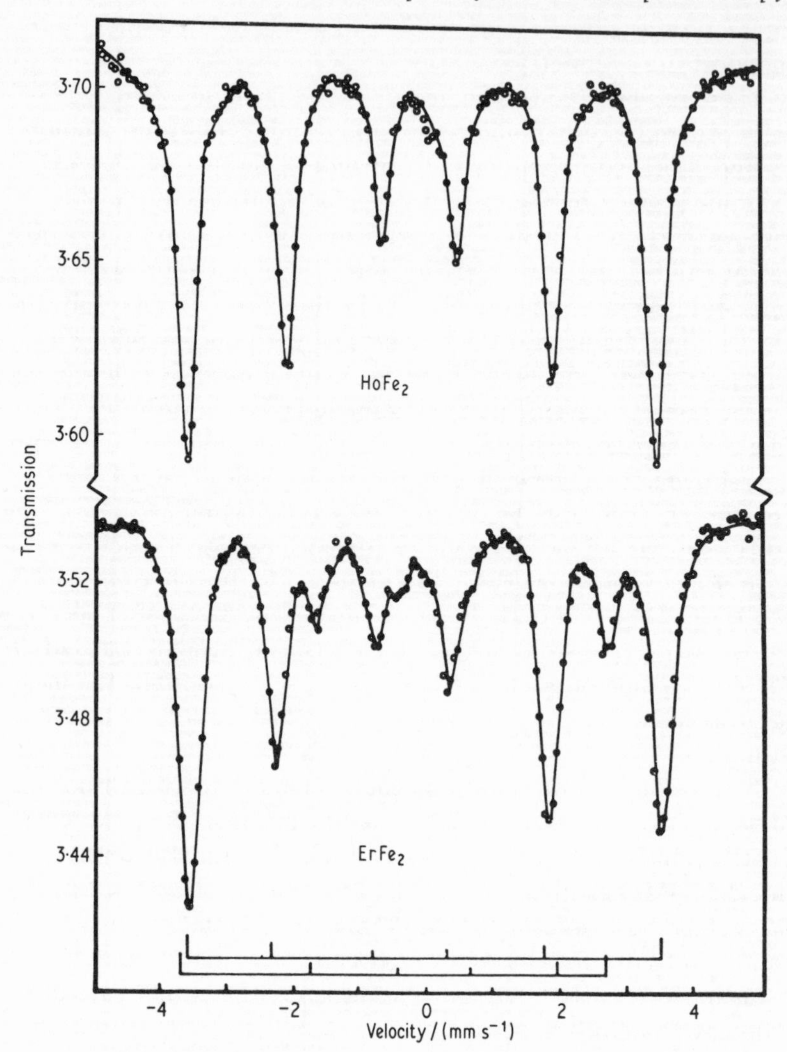


Fig. 8.3 The $^{57}Fe$ Mössbauer spectra at 77 K of $HoFe_2$ and $ErFe_2$ showing that only one iron site exists in the former, but two in the latter because of a different direction of the magnetic axis. ([6], Figs. 1, 3).

of the gold 6*s*-electrons in the alloy. This is additional to the *short-range* polarization caused by an immediately neighbouring manganese 3*d*-moment which gives the extra flux density of about 20 T.

On the other hand, in the intermetallic compounds $Au_4Mn$,

$Au_3Mn$, $Au_5Mn_2$ and $Au_2Mn$, the observed magnetic flux density is in direct proportion to the number and magnitude of the neighbouring Mn $3d$-moments, and clearly in this case the long-range polarization of the conduction electrons is less significant and is dominated by the short-range polarization.

$Au_2Mn$ has a layered body-centred tetragonal lattice and orders antiferromagnetically below 363 K. Each gold site has axial symmetry along the c axis and lies between a plane of gold and a plane of manganese atoms (see Fig. 8.4). The $^{197}Au$ resonance at 4.2 K shows a magnetic hyperfine splitting of the $\frac{1}{2} \rightarrow \frac{3}{2}$ transition with a large magnetic flux density of 157 T. The spectrum in Fig. 8.5 shows eight lines because of a substantial quadrupole (E2) admixture in the radiation which allows transitions with $\Delta m_z = \pm 2$ to take place [9]. The spectrum is asymmetric because of a quadrupole interaction. The solid lines represent the calculated spectrum assuming $V_{zz}$ (which must be along $c$ from the crystal symmetry) to be perpendicular to the magnetic axis. The transferred hyperfine field at the gold will be parallel or antiparallel to the $3d$-moment on the manganese which therefore must also be normal to the $c$ axis.

$AuMn_2$ has the same lattice but with the atoms reversed. Each gold layer lies between two manganese layers, and curiously the structure is not magnetically ordered at 4.2 K. Both $Au_2Mn$ and $AuMn_2$ show large quadrupole effects, the values of $e^2qQ_g$ being 2.9 and 4.64 mm $s^{-1}$ respectively [8]. From Fig. 8.4 it can be seen that the Au environment will be considerably distorted in both instances. The near-neighbours comprise a flattened cube with 4 Mn and 4 Au on opposite faces in $Au_2Mn$, and 8 Mn atoms in $AuMn_2$. The larger value of $e^2qQ$ in the second case shows that it is the manganese and not the gold neighbours which are largely responsible for the presence of the electric field gradient.

In the whole series of Au/Mn alloys and compounds there is an almost linear increase in the $^{197}Au$ chemical isomer shift as the gold content decreases. The total change of 6.6 mm $s^{-1}$ is much more than can be explained by the compression of the gold electrons by the decrease in cell volume. The most likely explanation [8] is a transfer of electrons from the transition metal to the $6s$-band of gold amounting to about 0.7 $e^-$ per atom.

The effects of compositional variation are particularly marked in the alloys based on the ordered body-centred cubic AuMn struc-

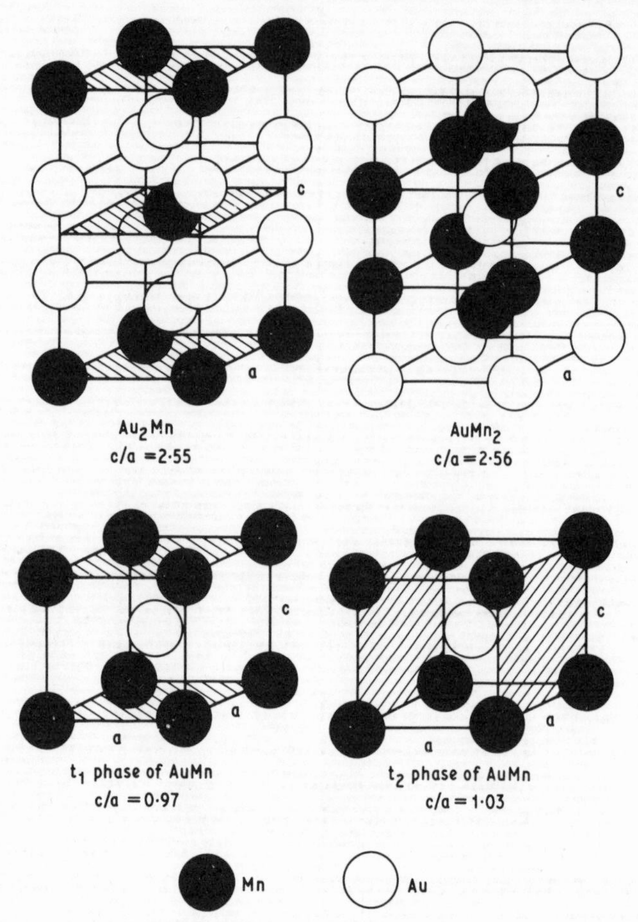


Fig. 8.4 The structures of $Au_2Mn$, $AuMn_2$, and the $t_1$ and $t_2$ phases of AuMn. The manganese spins are aligned parallel within any shaded plane, and antiparallel in adjacent planes (ignoring any spiral effects) to give an antiferromagnetic structure. $AuMn_2$ is non-magnetic.

ture (40–53 at.% gold). Typical $^{197}$Au spectra at 4.2 K are shown in Fig. 8.6. Upon cooling, the cubic high-temperature phase transforms initially to a tetragonally distorted $t_1$ phase ($c/a < 1$), but for less than 50 at.% Au there is a further phase transition to a second tetragonal phase, $t_2$ ($c/a > 1$), without any change in cell volume. Both tetragonal phases are antiferromagnetically ordered with alternating ferromagnetic sheets of Mn atoms perpendicular to a short axis (see Fig. 8.4). In the ordered 50 at.% alloy the

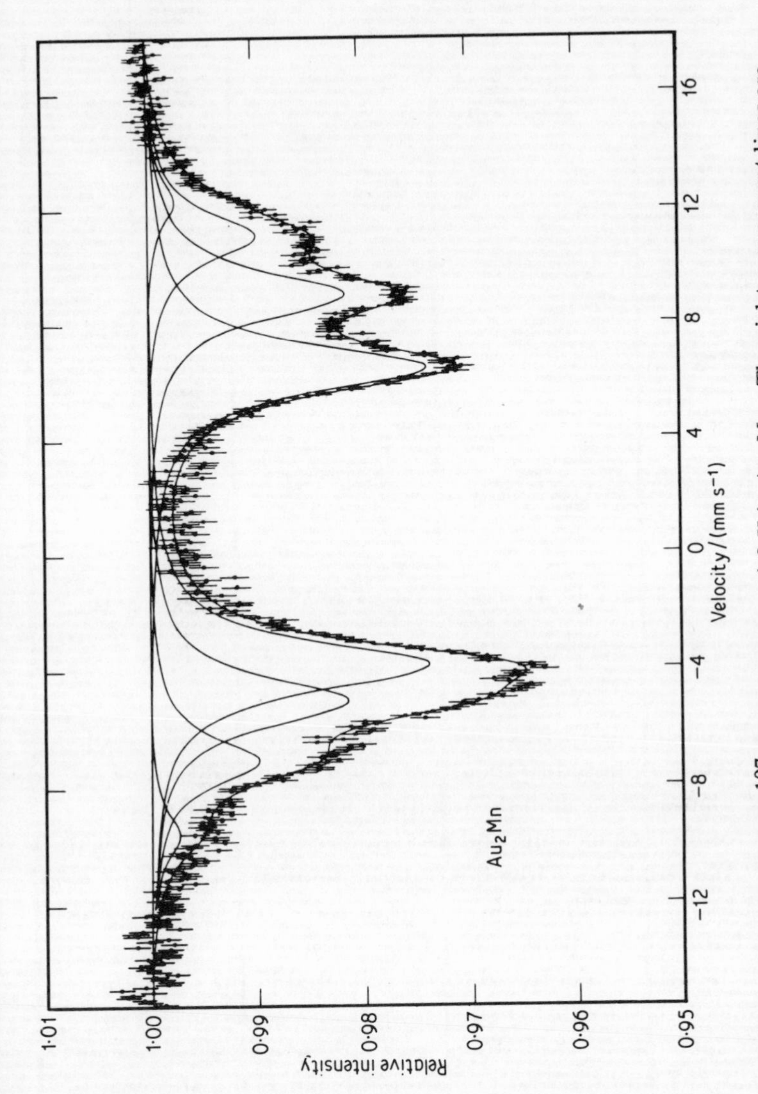


Fig. 8.5 The 77.34-keV 197Au resonance at 4.2 K in $Au_2Mn$. The eight component lines are from an $I_g = \frac{3}{2} \rightarrow I_e = \frac{1}{2}$ transition with E2/M1 mixing. The magnetic hyperfine splitting appears asymmetric because of a small quadrupole interaction. ([9], Fig. 2).

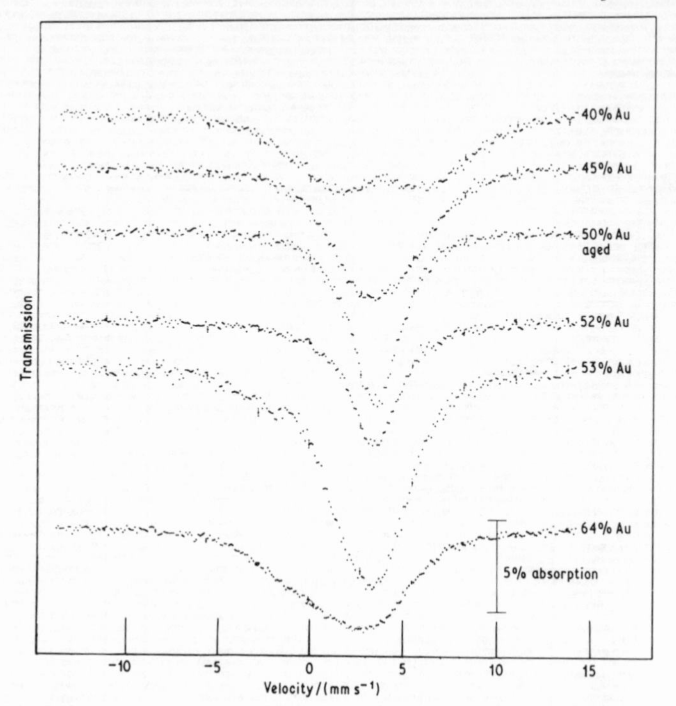


Fig. 8.6 The 77.34-keV $^{197}$Au spectra at 4.2 K of Mn/Au alloys close to the composition AuMn. Note the line broadening upon departure from stoichiometry in either direction. ([8], Fig. 3).

gold environment comprises eight Mn neighbours of which four have their spins opposed to the other four (being in different magnetic planes). The spin-polarization effects at the Au atom are therefore self-cancelling, and no field is expected or observed in the $^{197}$Au resonance below the ordering temperature. For a gold content of $<50$ at.%, the environment of each gold atom remains at 8 Mn, but the placement of Mn atoms in gold sites will disturb the magnetic arrangement, and this is reflected in the magnetic broadening of the Mössbauer spectrum. For a gold content of $>50$ at.%, some of the gold atoms have only seven nearest neighbour Mn atoms, and the resultant spin-imbalance will also result in a magnetic flux density at the gold nucleus. Thus a departure from stoichiometry in either direction results in some magnetic polarization of the gold 6*s*-electrons and a broadening of the $^{197}$Au resonance.

A further example of order-disorder effects is given by the Fe-Al system which maintains a body-centred cubic lattice from 0 to 54 at.% Al but tends to produce ordered structures near compositions corresponding to $Fe_3Al$ and FeAl. The degree of order-disorder is variable according to the thermal and mechanical treatment of the alloy [10]. The $^{57}Fe$ Mössbauer resonances of the ordered and disordered forms differ because of the changes in the near-neighbour environment of the iron. For example an alloy corresponding to FeAl and ordered by thermal annealing shows a narrow single-line absorption corresponding to the unique non-magnetic iron site in the ordered CsCl structure (illustrated in Fig. 8.7). The crushed alloy shows significant line broadening and in particular a gross broadening in the wings of the resonance characteri-

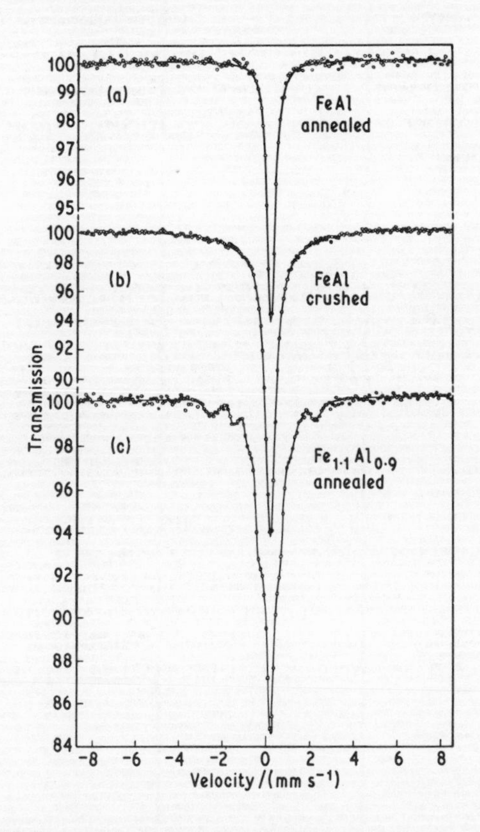


Fig. 8.7 The $^{57}Fe$ spectra at 4.2 K of (a) ordered FeAl (b) crushed FeAl (c) $Fe_{1.1}Al_{0.9}$. ([10], Fig. 2).

stic of an inhomogeneous magnetic hyperfine interaction. Plastic deformation of the body-centred cubic lattice will generate slip planes, giving antiphase boundaries or stacking faults containing an excess of like-atom pairs. This increase in the number of contiguous iron atoms is responsible for the magnetic behaviour.

An indirect verification of this interpretation can be obtained by examining a non-stoichiometric ordered alloy of the type $Fe_{1+x}Al_{1-x}$. Each iron atom on an Al site has eight Fe nearest neighbours, while each Fe site has dominantly between 6 and 8 Al nearest neighbours. The spectrum for an alloy with $x = 0.1$ is shown in Fig. 8.7. The magnetic area in the wings of the spectrum (about 9% of the total area) corresponds closely to the fraction of iron on Al sites which are thus centred in an $Fe_9$ cluster. The eight neighbouring iron atoms on Fe sites are less strongly coupled, but are nevertheless affected as seen by the broadening of the central component. The ferromagnetic behaviour of nearly stoichiometric FeAl is therefore determined by the presence of magnetic clusters, the number of which may be varied by appropriate treatment of the alloy.

## References

[1] Cordey Hayes, M. and Harris, I. R. (1967) *Phys. Letters,* **24A**, 80.
[2] Vincze, I. and Campbell, I. A. (1973) *J. Phys. F,* **3**, 647.
[3] Cranshaw, T. E. (1972) *J. Phys. F,* **2**, 615.
[4] Stearns, M. B. (1963) *Phys. Rev.,* **129**, 1136.
[5] Tansil, J. E., Obenshain, F. E. and Czjzek, G. (1972) *Phys. Rev. B,* **6**, 2796.
[6] Bowden, G. J., Bunbury, D. St. P., Guimarães, A. P. and Snyder, R. E. (1968) *J. Phys. C,* **1**, 1376.
[7] Bowden, G. J. (1973) *J. Phys. F,* **3**, 2206.
[8] Longworth, G. and Window, B. (1971) *J. Phys. F,* **1**, 217.
[9] Thompson, J. O., Huray, P. G., Patterson, D. O. and Roberts, L. D. (1968) In *Hyperfine Structure and Nuclear Radiations,* ed. E. Matthias and D. A. Shirley, p. 557, North Holland, Amsterdam.
[10] Wertheim, G. K. and Wernick, J. H. (1967) *Acta Met.,* **15**, 297.

CHAPTER NINE

# Analytical Applications

The preceding chapters have been chiefly concerned with the Mössbauer spectra of single-phase materials containing a small number of distinct atomic sites for the resonant atom. The more complicated behaviour seen in non-stoichiometric phases has been found to relate to the statistical variations in the near-neighbour environment. However, there are other solid-state systems which are either heterogeneous multi-phase materials, or have a complicated structure in which the atomic environment is variable and may even be unknown. Typical examples in this category are supported catalysts and ion-exchange resins. Nevertheless, Mössbauer spectroscopy can be invaluable as a means of investigating such systems. The sensitivity of the resonance to changes in oxidation state and site symmetry, and the essentially *non-destructive* nature of the technique can be used to good effect. This chapter describes a miscellaneous selection of analytical applications grouped under three general headings, but in no way attempts to be comprehensive.

## 9.1 Chemical analysis

It would be misleading to imply that Mössbauer spectroscopy is ideally suited to chemical analysis in general. Although in principle the intensity of an absorption can be related exactly to the amount of the resonant isotope in the absorber, the problems involved in correcting for the effects of the radiation background, and the need to know the recoilless fraction accurately, tend to make the experimental result subject to errors larger than those obtained by more conventional analysis. However, there are many examples where the elemental analysis cannot fully characterize a compound because two or more elements have alternative oxidation states.

An interesting example is provided by a blue iron-titanium anhydrous double sulphate which approximates in composition to the formula $FeTi(SO_4)_3$ with iron in the $Fe^{2+}$ oxidation state [1]. Repeated attempts to prepare a stoichiometric compound have failed; in every case the product appears to be a single phase material, but the chemical analyses always show an excess of iron and a significant but variable proportion of $Fe^{3+}$. The Mössbauer spectra at 295 K are comparatively simple, with a quadrupole doublet from $Fe^{2+}$ ($\Delta = 0.98$, $\delta = 1.26$ mm s$^{-1}$ with respect to Fe metal) and a broad singlet from $Fe^{3+}$ ($\delta = 0.64$ mm s$^{-1}$). However, in every preparation the proportion of iron in the $Fe^{3+}$ oxidation state, as estimated approximately from the relative absorption areas, is found to be appreciably higher than indicated by chemical analysis (the figures for the sample closest to stoichiometry being 28% by Mössbauer spectroscopy as opposed to 5.5% by analysis).

The formula of the hypothetical ideal compound is $(Fe^{2+})(Ti^{4+})(SO_4^{2-})_3$, and indeed the ratio of Fe + Ti to sulphate is always found to be very close to 2:3. In order to incorporate the excess of iron, a formulation of $(Fe^{2+}_{1-x}Fe^{3+}_x)(Fe^{3+}_xTi^{4+}_{1-x})(SO_4^{2-})_3$ must be adopted, which is then consistent with the analyses, but not with the Mössbauer data. The solution to the problem is to propose the existence of a proportion of $Ti^{3+}$ in the solid phase. The new formulation is $(Fe^{2+}_{1-x-y}Fe^{3+}_{x+y})(Fe^{3+}_xTi^{3+}_yTi^{4+}_{1-x-y})(SO_4^{2-})_3$, which in the limit of apparent stoichiometry becomes $(Fe^{2+}_{1-y}Fe^{3+}_y)(Ti^{3+}_yTi^{4+}_{1-y})(SO_4^{2-})_3$. The unstable $Ti^{3+}$ ion will reduce $Fe^{3+}$ to $Fe^{2+}$ on dissolution, so that the solution analysis gives a low value for the $Fe^{3+}$ content of the solid. Unfortunately, the crystal structure of this interesting compound has yet to be determined.

A second example is provided by the solid-solution $SrFe_xRu_{1-x}O_{3-y}$ derived from $SrFeO_3$ and $SrRuO_3$, both of which have the perovskite lattice with $M^{4+}$ cations but are antiferromagnetic and ferromagnetic respectively [2]. The perovskite lattice is often oxygen deficient, and in characterizing the solid-solution phase it is important to know the oxygen content. Unfortunately, for this combination of elements it is difficult to obtain analyses of sufficient accuracy to determine the value of $y$. However, the oxygen deficiency can be determined indirectly from the $^{99}Ru$ and $^{57}Fe$ Mössbauer spectra which give the oxidation states of the cations. Thus substitution of iron in $SrRuO_3$ takes place entirely as $Fe^{3+}$ and oxygen deficiency increases, so that compositions such as $SrFe^{3+}_{0.1}Ru^{4+}_{0.9}O_{2.95}$ and $SrFe^{3+}_{0.2}Ru^{4+}_{0.8}O_{2.90}$ are obtained. From about $x = 0.3$ it appears that the increase in oxygen deficiency is halted by the introduction of some $Ru^{5+}$, and at $x = 0.4$ the composition is approximately $SrFe^{3+}_{0.4}Ru^{4+}_{0.39}Ru^{5+}_{0.21}O_{2.91}$. At $x = 0.5$ the oxygen deficiency appears to be minimal with the unexpected composition of $SrFe^{3+}_{0.5}Ru^{5+}_{0.5}O_3$. Such detailed information could not be obtained by conventional analysis.

One rarely reported but very useful application of the Mössbauer spectrum is as a 'fingerprint' to check the efficacy of a chemical preparation. This is particularly true for co-ordination and organometallic compounds of iron and tin, where unwanted products are often easily detected by the presence of additional components in the Mössbauer spectrum. A preliminary examination in this way can be carried out quickly, and is completely non-destructive of the material.

This non-destructive aspect of the technique can be particularly useful. Furthermore, almost any size or shape of object can be examined if the experiment is carried out in a scattering mode. One unusual application is in the classification of works of art. Many of the yellow and brown pigments used in oil paintings such as ochres, siennas and umbers contain a substantial proportion of iron, mainly in the form of the magnetic oxides $\alpha$-$Fe_2O_3$ and $\alpha$-FeOOH. However, the naturally occuring deposits from which they were originally derived are variable in quality and average particle size. A systematic survey [3] of iron pigments has shown that the room-temperature Mössbauer spectrum is often characteristic, with a degree of superparamagnetic collapse of the hyperfine pattern which relates directly to the particle size. Pigments have been successfully identified in 18th-century oil paintings, and a systematic study of the work of a particular artist should prove valuable in the corroboration of doubtfully attributed paintings.

The scattering method is also applicable to analysis of terracotta statuary whose colour derives mainly from the presence of iron [3]. Typical spectra from two terracotta works are shown in

Fig. 9.1. The appearance of magnetic components in the sample TC-19 has been established to be the result of firing the clay to a higher temperature than was the case for TC-3. Individual craftsmen undoubtedly varied in their methods of furnace handling, and this distinction is also useful in the attribution of works to a particular artist. Fake statues made from different material can be readily distinguished from the genuine articles without damage to either.

The application of scattering techniques in the study of surface effects is considered in a separate section later in the chapter.

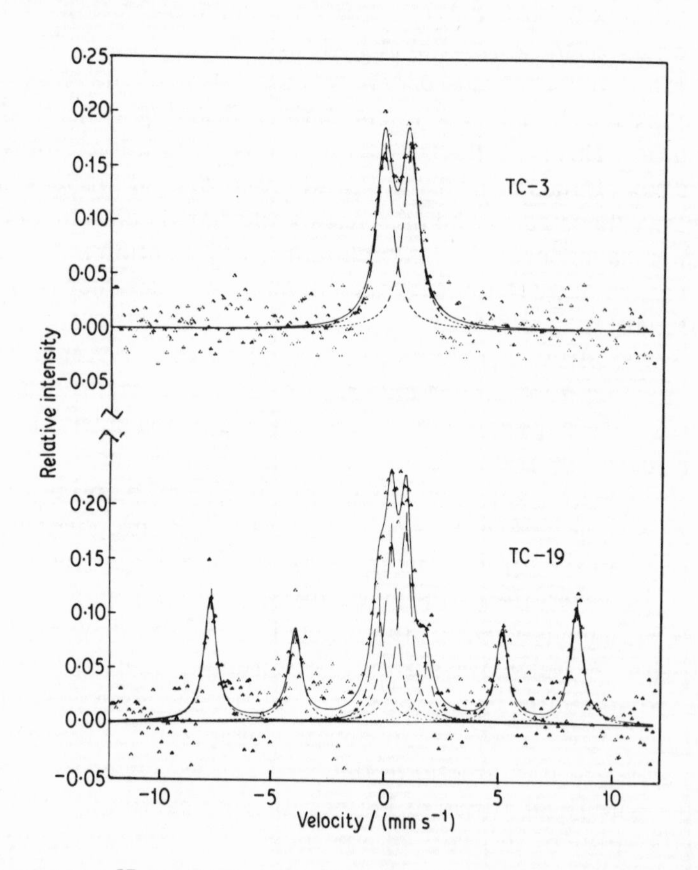


Fig. 9.1 The $^{57}$Fe Mössbauer spectrum in a scattering geometry from two samples of terracotta statuary. The magnetic components in the spectrum from TC-19 establish that it was fired to a higher temperature than TC-3. ([3], Figs. 20, 21).

## 9.2 Silicate minerals

The many naturally occurring silicate minerals are all structurally derived from the tetrahedral bonding of silicon to oxygen. The discrete orthosilicate anion, $SiO_4^{4-}$, is found in only a small proportion of them. In the vast majority the $SiO_4$ tetrahedra are joined by oxygen sharing into infinite chains or sheets. Examples are the pyroxenes with single-strand chains of composition $(SiO_3^{2-})_n$, the amphiboles with double-strand cross-linked chains of composition $(Si_4O_{11}^{6-})_n$ and the micas with infinite sheets of composition $(Si_2O_5^{2-})_n$. The silicon-oxygen anionic groupings are held together by metal cations between the chains or sheets in sites with usually four or six co-ordination to oxygen. The strength and rigidity of the silicon-oxygen lattice allow an unusually large variation in the nature of the cations which serve primarily to produce electroneutrality. Partial replacement of silicon by aluminium results in the closely related feldspars, zeolites and ultramarines. The silicates therefore show a greater degree of compositional variation than is found for example in simple oxides.

Many of the silicates contain a substantial proportion of iron as $Fe^{2+}$ and $Fe^{3+}$ cations, and therefore give a good $^{57}Fe$ Mössbauer absorption. The number of distinct cation sites is usually limited to between one and four. The distribution of the various cations within these sites is not completely random. A substantial degree of ordering is often found, determined largely by the differing sizes of the lattice sites and of the cations which occupy them. To determine quantitative site populations is not an easy matter, and Mössbauer spectroscopy represents one of the most useful methods for this purpose.

The effects on the Mössbauer spectrum of the immediate site co-ordination in an iron oxide are described in Chapter 7, and these principles still hold in the silicates. The main difference lies in that the somewhat random occupation of the near-neighbour cation sites by widely different cations tends to produce a noticeable broadening of the resonance lines. Magnetic effects are almost unknown, and any deductions must be made from the chemical isomer shift, quadrupole splitting, and intensity of absorption. The large difference in chemical isomer shift between $Fe^{2+}$ and $Fe^{3+}$ enables the approximate proportions of the two oxidation states to be determined quite easily from the relative areas of the absorption lines. The recoilless fractions at different sites are usually similar, and the error introduced by assuming them to be equal is small. The quadrupole splitting of an $Fe^{3+}$ component is generally small, and it is difficult to distinguish multiple site symmetries for this cation. However, the comparatively large splitting found for an $Fe^{2+}$ site and its sensitivity to environment usually results in a resolved fine structure, leading directly to an estimate of the relative site occupancy by the $Fe^{2+}$ ions. Both the chemical isomer shift and quadrupole splitting of an $Fe^{2+}$ ion decrease as the co-ordination decreases from eight to six to four. As a result it is possible to make deductions concerning site symmetry and occupation in silicates of unknown structure. Many mineral samples are strongly textured (particularly the amphiboles and micas) and cannot be made into an absorber without some residual orientation. As a result the two components of a quadrupole doublet may have an unequal intensity.

These general remarks can be illustrated by the spectra in Fig. 9.2. The mineral howieite is believed to have a chain structure and

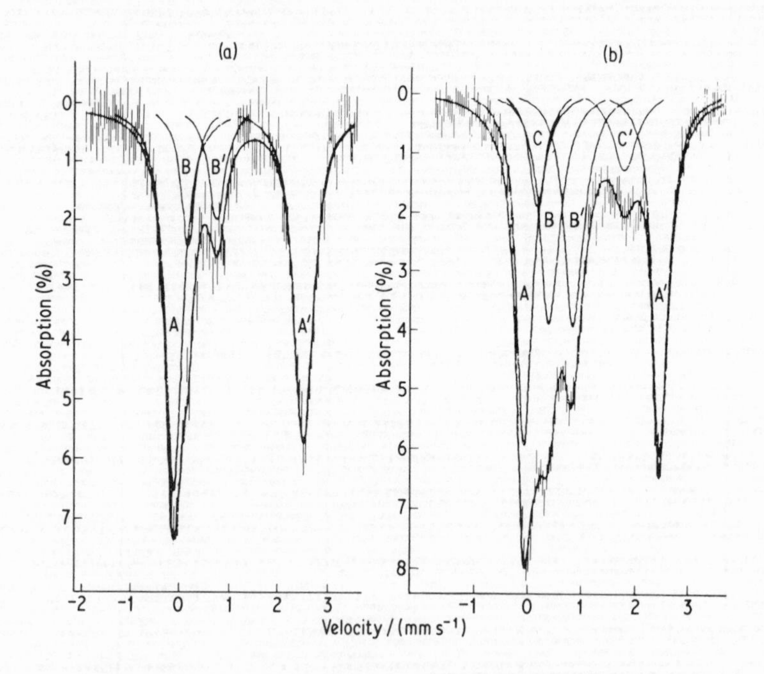


Fig. 9.2 The room temperature $^{57}Fe$ Mössbauer spectra of (a) howieite and (b) deerite. ([4], Fig. 1).

has the approximate composition $NaFe_7^{2+}Mn_3Fe_2^{3+}Si_{12}O_{31}(OH)_{13}$. The room-temperature spectrum [4] shows two quadrupole doublets, that with the larger splitting (AA′) having the high chemical isomer shift and large quadrupole splitting (see Table 9.1) typical of high-spin $Fe^{2+}$ ions in octahedral co-ordination. The inner doublet (BB′) is typical of high-spin $Fe^{3+}$ ions, also in octahedral co-ordination. The mineral deerite has the approximate composition $Fe_{13}^{2+}Fe_7^{3+}Si_{13}O_{44}(OH)_{11}$. The spectrum shows three quadrupole doublets. AA′ arises from $Fe^{2+}$ ions in distorted six co-ordination, BB′ from $Fe^{3+}$ in distorted six co-ordination, while the CC′ lines with a significantly lower shift than AA′ but still within the range for high-spin $Fe^{2+}$ are clearly from a four co-ordinated tetrahedral site. Note that the two components of each doublet are not necessarily equal in intensity because of texture in the absorber.

A better known class of silicates is the pyroxenes which contain single-stranded chains of $SiO_4$ groups of overall composition $(SiO_3^{2-})_n$. The general formula can be written as $ABSi_2O_6$ where A and B represent distinct cation sites known as $M_1$ and $M_2$ respectively. The $M_1$ sites have a near-regular six co-ordination to oxygen and are usually occupied by small cations such as $Mg^{2+}$, $Al^{3+}$, $Mn^{2+}$, $Fe^{3+}$ or $Na^+$. The $M_2$ site is grossly distorted, and being larger tends to be occupied by $Na^+$, $Mg^{2+}$, $Ca^{2+}$, $Mn^{2+}$ and $Fe^{2+}$. All pyroxenes do not have a common space group; for example pigeonite has the space group $P2_1/c$, $CaMgSi_2O_6$ and $CaFeSi_2O_6$ belong to $C2/c$, and $Mg_2Si_2O_6$ to Pbca, but the small differences involved have only a minor influence on the Mössbauer spectra.

In all the pyroxenes the $Fe^{2+}$ cations at the more distorted $M_2$ sites show a *smaller* quadrupole splitting than at the $M_1$ sites. This

Table 9.1 Mössbauer parameters for howieite and deerite at room temperature

| Compound | Site | $\delta/(mm\ s^{-1})$* | $\Delta/(mm\ s^{-1})$ |
|---|---|---|---|
| Howieite | A | 1.18 | 2.81 |
| | B | 0.40 | 0.59 |
| Deerite | A | 1.13 | 2.57 |
| | B | 0.45 | 0.59 |
| | C | 1.02 | 1.40 |

* relative to Fe metal.

is probably because of a large 'lattice' contribution to the electric field gradient tensor which is opposite in sign to the 'valence' term from the $3d^6$ configuration (see p. 103). The $^{57}Fe$ spectra comprise two superimposed doublets which are partially resolved (ignoring for the present a possible third contribution if $Fe^{3+}$ ions are also present; some examples of spectra from orthopyroxenes are shown in Fig. 9.4).

It is convenient to limit further discussion to the orthopyroxenes, which have a composition close to the system $MgSiO_3$-$FeSiO_3$. The $Fe^{2+}$ cations prefer to occupy the $M_2$ sites, but considerable disorder can be induced by heating to 1270 K or above. Annealing at lower temperatures causes a preferential occupation of the $M_2$ sites by $Fe^{2+}$ which increases with decreasing temperature until at about 750 K the ionic diffusion is apparently frozen [5, 6]. If the relative occupations of the $M_1$ and $M_2$ sites by $Fe^{2+}$ are $x_1$ and $x_2$ respectively, then a parameter $k$ can be defined by

$$k = \frac{x_1(1 - x_2)}{x_2(1 - x_1)}$$

For a system *at equilibrium*, $k$ corresponds to a true equilibrium constant for the reaction

$$\mathrm{Fe(M_2) + Mg(M_1) \rightleftharpoons Fe(M_1) + Mg(M_2)}$$

and can be related to the free-energy difference $\Delta G_E^\circ$ by

$$\Delta G_E^\circ = -RT \ln k$$

If $\Delta G_E^\circ$ is independent of temperature, $k$ should decrease with decreasing temperature (i.e. $x_1$ decreases, $x_2$ decreases).

The values of $x_1$ and $x_2$ for a wide range of natural orthopyroxenes are plotted in Fig. 9.3. The open circles correspond to natural samples re-heated at 1270 K until equilibrium had been reached and then rapidly quenched to preserve the disorder. The solid curve represents the theoretical prediction for $k = 0.235$ ($\Delta G_E^\circ = 15.30$ kJ mol$^{-1}$), and shows that the relationship holds for $x_1 < 0.6$. The filled circles represent metamorphic and plutonic rocks which have cooled slowly and have therefore reached a

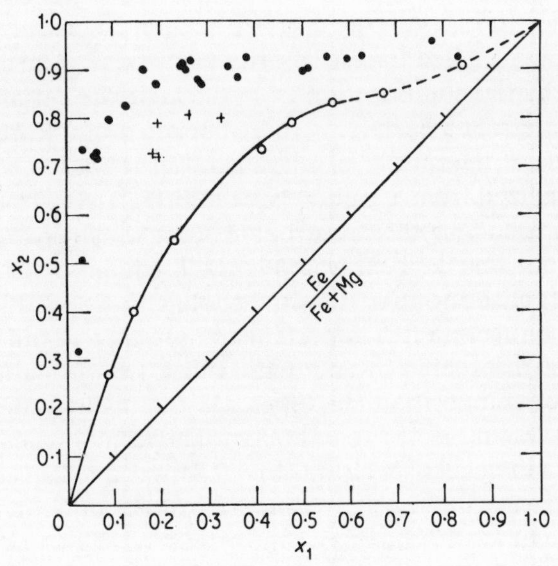


Fig. 9.3 A plot of the measured site occupancies $x_1$ and $x_2$ of the $M_1$ and $M_2$ sites in a selection of natural orthopyroxenes. The open circles represent natural samples reheated to 1270 K and quenched. The filled circles represent natural slowly cooled metamorphic and plutonic rocks. The crosses represent quickly cooled volcanic rocks. ([6], Fig. 1).

final equilibrium at about 750 K with $k = 0.028$. Note the preferential occupation of $M_2$ sites by $Fe^{2+}$. The crosses represent volcanic rocks which have cooled more quickly and have therefore retained a proportion of non-equilibrium disorder.

The pyroxenes are now known to comprise a major proportion of the rocks and soil on the lunar surface (together with other silicates such as olivine and plagioclase, opaque minerals such as ilmenite, troilite, metallic iron/nickel alloy and spinel oxides, and vitrified material). Being non-destructive, Mössbauer spectroscopy has proved to be a useful method of studying the material returned to earth by the American 'Apollo' and Russian 'Luna' spacecraft. In some instances it has proved possible to obtain orthopyroxene separates from lunar rocks. Such a sample separated from an igneous rock returned by the 'Apollo 14' mission (sample 14310, 116-P1) has a composition of $Mg_{1.40}Fe_{0.48}Ca_{0.12}Si_2O_6$ [7]. The $M_1$ and $M_2$ site occupancies have been determined to be $x_1 = 0.076$ and $x_2 = 0.398$, leading to a value for $k$ of 0.097. As may be seen by reference to Fig. 9.3, the degree of cation disorder is higher than in slowly cooled terrestrial orthopyroxenes. The cooling rate of this rock after solidification from the melt was certainly faster than expected for a deep-lying rock, but probably slower than one would find for the surface of a lava flow. The rock is believed to have crystallized from a molten pocket in the ejecta blanket at *Fra Mauro* following the catastrophic event which created *Mare Imbrium*, but was then excavated in a subsequent event without substantial reheating *after* its initial cooling to below 750 K.

In this connection it is important to note that an ordered pyroxene can be disordered by an intense shock [8]. A natural orthopyroxene of composition $Mg_{1.72}Fe_{0.28}Si_2O_6$ (highly ordered with $k = 0.041$) was subjected to a shock of between 900 and 1000 kbar. The Mössbauer spectrum (Fig. 9.4) was unchanged apart from an increase in the relative intensity of the outer doublet to give a value of $k = 0.27$. One of the effects of the shock is to cause a short-term heating to as high as 1650 K. The result is a rapid diffusion of cations which are then quenched into a disordered state. It therefore follows that intensively shocked meteorite or lunar basalt will have lost any thermal record held prior to the catastrophic event. It is also clear that lunar rock 14310,116 cannot have been intensively shocked during excavation to the surface by a meteorite impact.

## 9.3 Surface chemistry

There are several important aspects of chemistry which are intimately concerned with the properties of surfaces, including such topics as corrosion, adsorption, and catalysis at solid-liquid and solid-gas interfaces. In propitious cases it is still possible to make Mössbauer measurements in conventional transmission geometry, but for surface studies in general it is often more appropriate to use a scattering experiment. The detected radiation can then be either the resonantly scattered $\gamma$-ray, or alternatively the X-rays or electrons produced by internal conversion. The final choice is dictated partly by the properties of the $\gamma$-decay, and partly by the nature of the surface.

For example, the detection of resonant scattering from $^{57}Fe$ in steel plates is best monitored by using the 6.3-keV X-rays rather than the primary $\gamma$-ray, and provides a means of studying corro-

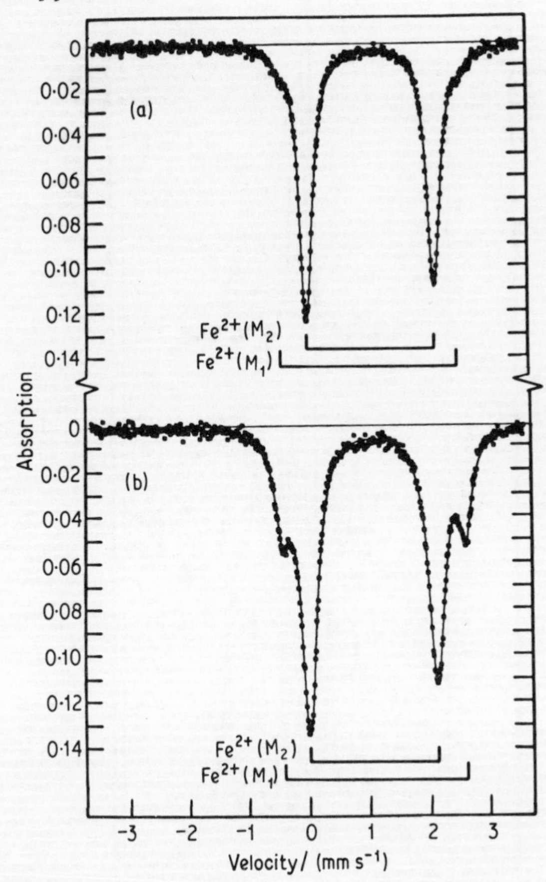


Fig. 9.4 The $^{57}$Fe Mössbauer spectrum at 77 K of an orthopyroxene of composition $Mg_{1.72}Fe_{0.28}Si_2O_6$ (a) in natural form and (b) after intensive shock. ([8], Fig. 1).

sion [9]. The spectrum for scattering from a steel plate exposed to fumes of HCl in air is shown in Fig. 9.5a. The dominant doublet in the spectrum can be identified from the chemical isomer shift and quadrupole splitting as being that of $\beta$-FeOOH and is distinct from the spectra of the other oxides and oxyhydroxides of iron. The intensity of the scattering can be related to the depth of the surface layer by evaluation of the appropriate scattering integral, and in this case gives an approximate thickness of about $2 \times 10^{-3}$ cm. The additional weak components on the baseline are the magnetic hyperfine pattern of the underlying steel. Fig. 9.5b

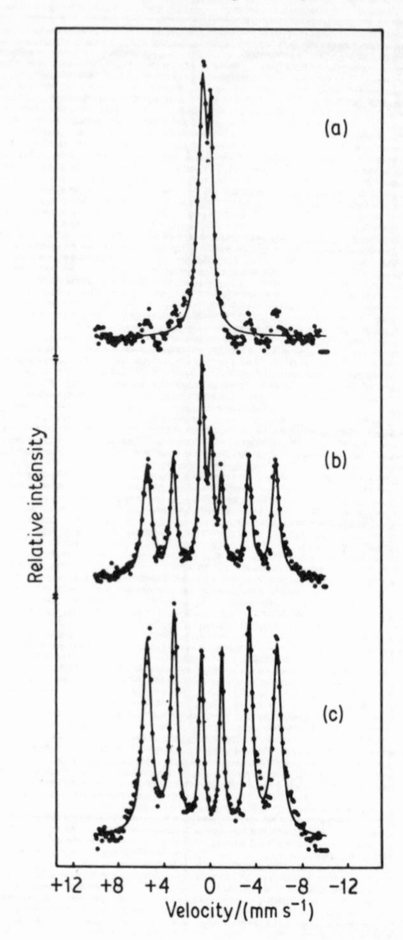


Fig. 9.5 The $^{57}$Fe Mössbauer spectrum at room temperature obtained by scattering from a corroded steel plate: (a) the corroded plate showing the strong central absorption from β-FeOOH; (b) as in (a) but with the rust scraped off; (c) as in (b) but with the surface milled clean. ([9], Fig. 2).

shows the spectrum recorded after attempting to scrape off the rust, and reveals that there is still a substantial rust residue. Spectrum (c) was obtained after milling 1 mm off the steel to give a fully cleaned surface. Those resonant nuclei close to the surface give a larger contribution to the observed spectrum than deeper-lying nuclei because the non-resonant attenuation of the X-rays increases with depth. The effective 'surface' is of the order of $10^{-3}$ cm.

In some ways a more valuable technique for surface studies is to record the back-scattering of the conversion electrons. The attenuation of electrons by a solid is much greater than for the $\gamma$-rays, so that it is only possible to record events which take place at a depth of less than about 400 Å ($4 \times 10^{-6}$ cm) from the immediate surface. A convincing demonstration of this method has been given in a study of the thermal oxidation of iron [10]. To improve the quality of the spectra the natural iron foils were electroplated with $^{57}Fe$ and annealed in hydrogen to form an isotopically enriched surface layer. Oxidation at 225°C for varying times showed the gradual formation of a surface layer of non-stoichiometric $Fe_3O_4$, and this is illustrated in Fig. 9.6.

The rate of growth of the surface layer (as measured by the change in the thickness $W$ as a function of time $t$) will follow a different rate law according to the mechanism of the corrosion [11]. For example a parabolic growth law of the form $W^2 = At + B$ is usually found when the movement of ions takes place through the lattice of the surface layer by diffusional migration (i.e. the cations move outwards or the anions inwards) and the resultant surface film is fairly uniform in thickness. Penetration of oxygen into the interior in a 'leakage' manner rather than by uniform diffusion results in a logarithmic law of the form $W = A \ln(Bt + C)$ and a less uniform layer. The intensity of the Mössbauer resonance lines from the growing surface layer in this experiment follow a logarithmic time dependence, consistent with the 'leakage' mechanism being dominant at 225°C. Spectrum (d) in Fig. 9.6 was obtained after oxidation for 120 minutes, and corresponds to a thickness of about 100 Å ($10^{-6}$ cm).

Oxidation at 350°C for five minutes gave a two-component surface layer in which additional components due to $\alpha$-$Fe_2O_3$ and $Fe_3O_4$ could be distinguished. The minimum thickness of an oxide coat which can be detected by this method is estimated to be about 40 Å.

A more sophisticated development of this technique is to incorporate a focussing $\beta$-spectrometer and to measure the exact energies of the electrons emitted from the surface. The energy of the electron is decreased with increasing depth of the emitting atom below the surface as the result of scattering. It is therefore possible to distinguish electrons originating in different sub-surface layers. This has been convincingly demonstrated [12] using a metallic tin plate exposed to bromine vapour. The Mössbauer

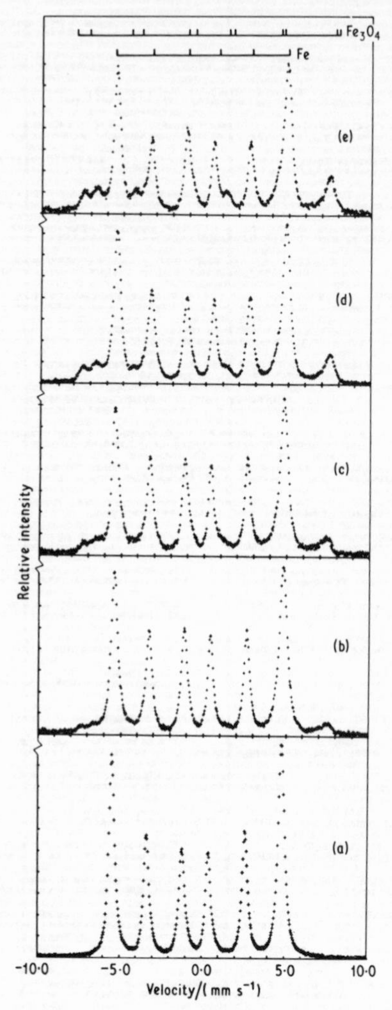


Fig. 9.6 The $^{57}$Fe Mössbauer spectra at room temperature (obtained by counting the conversion electrons emitted after resonant absorption in a scattering geometry) of an iron foil oxidized at 225°C for varying periods of time: (a) before oxidation (b) 5 minutes (c) 15 minutes (d) 120 minutes (e) 1000 minutes. ([10], Fig.4).

spectrum corresponding to a considerable depth below the surface is that of pure β-tin. Layers nearer to the surface show an increasing absorption due initially to $SnBr_2$ and ultimately to $SnBr_4$.

The study of adsorption at surfaces presents considerable problems, but is of immense importance with regard to catalysis and ion-exchange. One type of catalyst which has been extensively studied is made from α-$Fe_2O_3$ microcrystallites supported on high-area oxides. For example silica gel can be impregnated with ferric nitrate and then calcined in air to give an 'active' deposit on the $SiO_2$ matrix. This preparation can easily be carried out with enriched $^{57}Fe$, and the Mössbauer spectrum recorded by transmission methods [13].

There are two aspects to these investigations. Firstly, the properties of the catalyst are very dependent on the method of preparation and its resultant influence on the surface area and the particle size of the $Fe_2O_3$. The latter can be conveniently investigated using the $^{57}Fe$ Mössbauer spectrum because the magnetic hyperfine splitting of bulk antiferromagnetic α-$Fe_2O_3$ is partially or completely collapsed by superparamagnetic relaxation in the microcrystalline state (see p. 156). The temperature-dependence of the spectrum can therefore be easily related to the mean particle size.

Secondly, the adsorption of ions or radicals onto the surface of the iron oxide often has a marked effect on the Mössbauer spectrum, although it is not easy to determine why the changes take place. Adsorption of ammonia [14], water or $H_2S$ [15] completely changes the Mössbauer absorption; for example addition of water reduces the $Fe^{3+}$ quadrupole splitting from 2.24 mm $s^{-1}$ to 1.56 mm $s^{-1}$ at room temperature. All the iron atoms appear to be affected, and the exact nature and mode of action of the activated surface are not yet fully understood.

The investigation of zeolites by Mössbauer methods differs in that the iron is not incorporated into the support but is adsorbed onto the interior surface. The Mössbauer parameters of $Fe^{2+}$ or $Fe^{3+}$ cations in the zeolite can in principle be related to the symmetry of the adsorption sites in the aluminosilicate framework [16]. Once these sites have been characterized, it then becomes possible to undertake selective replacement adsorption with other species, and in this way to indirectly identify the adsorption sites for atoms and molecules containing only non-Mössbauer elements.

Where the adsorption is slow, the examination of a series of samples exposed to reagents for varying periods of time can be used for kinetic studies.

## References

[1] Gibb, T. C., Greenwood, N. N., Tetlow, A. and Twist, W. (1968) *J. Chem. Soc. (A),* 2955.
[2] Gibb, T. C., Greatrex, R., Greenwood, N. N. and Snowdon, K. G. (1974) *J. Solid State Chem.,* **14**, 00.
[3] Keisch, B. (1973) *Archaeometry,* **15**, 79.
[4] Bancroft, G. M., Burns, R. G. and Stone, A. J. (1968) *Geochim. Cosmochim. Acta,* **32**, 547.
[5] Virgo, D. and Hafner, S. S. (1970) *Amer. Mineral,* **55**, 201.
[6] Hafner, S. S. and Virgo, D. (1969) *Science,* **165**, 285.
[7] Schürmann, K. and Hafner, S. S. (1972) Proc. Third Lunar Sci. Conf., *Geochim. Cosmochim. Acta,* Suppl. 3, vol. **1**, 493.
[8] Dundon, R. W. and Hafner, S. S. (1971) *Science,* **174**, 581.
[9] Terrell, J. H. and Spijkerman, J. J. (1968) *Appl. Phys. Lett.,* **13**, 11.
[10] Simmons, G. W., Kellerman, E. and Leidheiser, H. (1973) *Corrosion,* **29**, 227.
[11] Davies, D. E., Evans, U. R. and Agar, J. N. (1954) *Proc. Roy. Soc.,* **225A**, 443.
[12] Bonchev, Zw., Jordanov, A. and Minkova, A. (1969) *Nuclear Instr. Methods,* **70**, 36.
[13] Hobson, M. C. and Gager, H. M. (1970) *J. Catalysis,* **16**, 254.
[14] Hobson, M. C. and Gager, H. M. (1970) *J. Colloid & Interface Sci.,* **34**, 357.
[15] Gager, H. M., Lefelhocz, J. F. and Hobson, M. C. (1973) *Chem. Phys. Letters,* **23**, 387.
[16] Delgass, W. N., Garten, R. L. and Boudart, M. (1969) *J. Phys. Chem.,* **73**, 2970.

CHAPTER TEN

# *Impurity and Decay After-effect Studies*

One of the first applications of Mössbauer spectroscopy was the study of the behaviour of an impurity atom in a metallic host matrix. The vibrational characteristics of such an atom are of fundamental importance to an understanding of lattice dynamics, and may be measured using the temperature dependence of the chemical isomer shift and recoilless fraction (see Chapter 6). There are however a number of other ways in which impurity effects can be used. In some instances the interest centres upon the impurity atom itself, as for example in semiconductors where the physical and chemical properties may be drastically altered by the nature of the impurity. In other cases the impurity atom is incorporated into the host with the intention of monitoring the properties of the latter. The incorporation of isotopically enriched impurities into metals, oxides and other ionic salts for use as Mössbauer absorbers has been extensively used for both purposes.

There is however a lower limit to the concentration of an impurity which can be used to obtain a good resonant absorption, and this is of the order of 0.1 at.%. If very dilute concentrations of impurity atoms are required, it may be better to use the alternative procedure of incorporating the radioactive precursor into the host matrix, and then comparing this *source* with a single-line reference absorber. [Note:- A resonant absorption occurring at a *positive* velocity with respect to a standard source represents an energy greater than the standard. If the roles are now reversed, a resonant emission occurring at *negative* velocity with respect to a standard absorber also represents a higher energy. For this reason a hyperfine emission spectrum is the mirror image of a corresponding absorption spectrum. The sign of the chemical isomer shift must therefore be reversed for comparison with conventional absorption data.]

However, the radioactive decay of the parent nucleus may cause drastic chemical modification of the environment of the daughter nucleus before the Mössbauer $\gamma$-emission takes place. If the host matrix is metallic, then the high mobility of the electrons ensures that any after-effects of the decay are obliterated within a time much less than the Mössbauer excited-state lifetime. Therefore the doping of metals with for example $^{57}Co$ or $^{119m}Sn$ can be regarded simply as a means of producing very low concentrations of iron or tin impurity atoms. If on the other hand the host matrix is an insulator, then after-effects may be observed. These will be discussed separately. The after-effects of nuclear decay may be studied advantageously by Mössbauer spectroscopy because the technique is specific to the site of the disturbance and gives information concerning the atomic environment within $10^{-7}$ s of the primary event.

## 10.1 Impurity doping

Impurity doping can be used most effectively in simple lattice compounds such as halides and oxides where the co-ordinate geometry at the impurity site can be easily established. An excellent example is provided by cadmium fluoride doped with $^{57}Fe$ [1]. $CdF_2$ is isostructural with fluorite, $CaF_2$. Each $Cd^{2+}$ ion is at the centre of a cube of $F^-$ anions, and each $F^-$ is tetrahedrally surrounded by $Cd^{2+}$ ions. The ionic radius of $Cd^{2+}$ is 1.07 Å (107 pm). However, the radius of high-spin $Fe^{2+}$ is only about 0.81 Å so that substitution of $Fe^{2+}$ on a $Cd^{2+}$ site will leave a significant amount of 'rattle-room'. The doping of iron into $CdF_2$ can produce impurities in both the $Fe^{2+}$ and $Fe^{3+}$ oxidation states, but with appropriate thermal treatment it is possible to obtain samples containing only $Fe^{2+}$ ions. The Mössbauer spectra at 296 K and 5 K of $CdF_2$ doped with enriched $^{57}Fe$ ($Cd_{0.99}Fe_{0.01}F_2$) are shown in Fig. 10.1. The single-line resonance at 296 K has a chemical isomer shift of +1.444 mm $s^{-1}$ (with respect to Fe metal at the same temperature) which is typical of the $Fe^{2+}$ ion in a site with

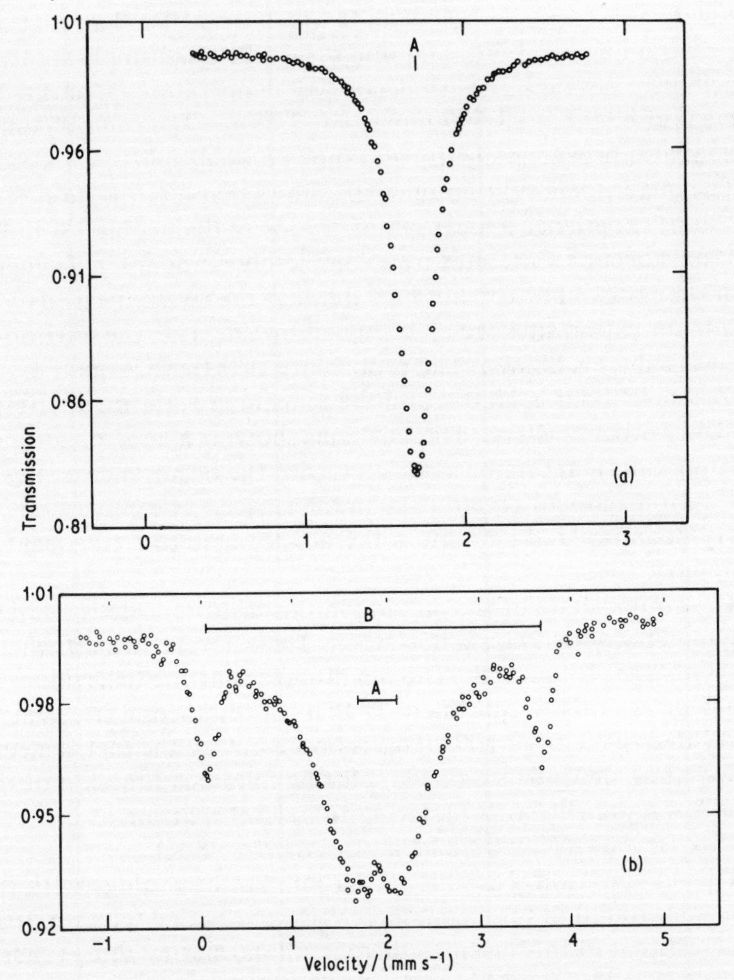


Fig. 10.1 The Mössbauer spectra of $Cd_{0.99}Fe_{0.01}F_2$ at (a) 296 K and (b) 5 K The weak doublet (B) seen at low temperatures corresponds to a near-neighbour $Fe^{2+}$-$Fe^{2+}$ cation pair, while the central resonance (A) corresponds to isolated $Fe^{2+}$ ions. ([1], Fig. 1).

little or no covalency. The lack of a quadrupole splitting shows that the impurity ion is occupying the cubic $Cd^{2+}$ site. The line-width of 0.26 mm $s^{-1}$ broadens slightly at lower temperatures (e.g. 0.34 mm $s^{-1}$ at 195 K) and becomes non-Lorentzian until at about 90 K a quadrupole-split doublet (B) comprising about 13% of the total area emerges from the wings of the central peak. This doublet has a chemical isomer shift which is smaller than that of the central line by about 0.04 mm $s^{-1}$, and a quadrupole splitting which increases rapidly with decreasing temperature from $\Delta = 1.735$ mm $s^{-1}$ at 89.9 K to 3.599 mm $s^{-1}$ at 2.08 K. The central component also broadens slightly and shows a small splitting (A).

The quadrupole-split doublet (B) shows that a proportion of the $Fe^{2+}$ ions experience a local symmetry lower than cubic. If the $Fe^{2+}$ ions occupy $Cd^{2+}$ sites by a completely random replacement, then the probability for any $Fe^{2+}$ ion that the twelve nearest-neighbour cation sites are all occupied by $Cd^{2+}$ (i.e. the environment is cubic) is 88.6%. The probability that there is one $Fe^{2+}$ nearest-neighbour is 10.0%, and the remaining 1.4% are distributed in higher-order clusters. The $Fe^{2+}$ ions showing a large quadrupole splitting can therefore be associated with the occurrence of $Fe^{2+}$–$Fe^{2+}$ near-neighbour cation pairs in the lattice. The cubic symmetry of an isolated $Fe^{2+}$ site is thereby lowered to tetragonal $C_{4v}$ symmetry.

The ground-state of the $3d^6$ ion in this distorted site is a level of $e_g$ symmetry with a $t_{2g}$ excited state. The $e_g$ level will split when the symmetry is lowered. The quadrupole-splitting behaviour is therefore essentially equivalent to that for tetrahedral site symmetry described on p. 107. This is borne out by the fact that the temperature dependence of the splitting follows a function of the type $\Delta = \Delta_0 \tanh(E_0/2kT)$, giving an $e_g$ level-splitting of $E_0 = 8.5 \times 10^{-3}$ eV ($E_0 = 0.82$ kJ $mol^{-1}$; $E_0/k = 98$ K).

The $3d^6$ $Fe^{2+}$ ion is a Jahn–Teller ion, but by analogy with the behaviour found in for example $FeV_2O_4$ and $FeCr_2O_4$ (see p. 163) it is clear that in this instance there is still no static Jahn–Teller distortion of the cubic $Fe^{2+}$ site at 5 K. The very small splitting of the central component can be explained satisfactorily as being due to random strains from distant defects distributed throughout the lattice.

The use of impurity doping to study the host matrix can be illustrated by data from single crystals of $MnF_2$, grown from a melt of $MnF_2$ containing isotopically enriched $FeF_2$ [2]. $MnF_2$ has the rutile ($TiO_2$) structure and is antiferromagnetic below 67.3 K. Substitution of $Fe^{2+}$ into the lattice at $Mn^{2+}$ sites causes a slight increase in the Néel temperature (by about 1 K), but otherwise has no catastrophic effects on the magnetic properties. At 4.2 K the $^{57}Fe$ spectrum shows magnetic hyperfine splitting with a flux density of 22.8 T. Because the absorber is a single crystal, the line intensities are angular dependent. As a result it is easy to show that the $^{57}Fe$ atomic moments which produce the flux density $B$, and hence the Mn moments of the host with which they are aligned, are parallel to the crystal $c$ axis. The magnetic axis in $MnF_2$ is therefore in the same direction as in isostructural $FeF_2$. The observed Mössbauer spectrum is complicated by a combination of magnetic and quadrupole interactions. Computation from the line positions reveals that the electric field gradient tensor has a principle value $V_{zz}$ which is perpendicular to $B$ and to the $c$ axis. The coupling constant $e^2qQ/2 = 2.80$ mm $s^{-1}$ and the asymmetry parameter $\eta = 0.5$. The tensor at the $Fe^{2+}$ site in $MnF_2$ is therefore very similar to that in $FeF_2$.

Apart from the previously mentioned application to the study of lattice dynamics, impurity doping in metals can also serve to examine electronic and magnetic structure. An interesting example is given by the systematic variations in the chemical isomer shift shown by $^{99}Rh$, $^{193}Os$ and $^{197}Pt$ impurities in the $3d$-, $4d$- and $5d$-transition metals [3]. These isotopes are the respective source isotopes for the 90-keV $^{99}Ru$, 73-keV $^{193}Ir$, and 77-keV $^{197}Au$ Mössbauer transitions. The observed shifts relative to metallic Ru, Ir or Au as appropriate are shown in Fig. 10.2. The following general trends may be observed:

(1) The value of $|\psi_s(0)|^2$ decreases nearly monotonically with an increase in the $d^n$ configuration of the host metal. [Note:- $\delta R/R$ is positive for all three isotopes]

(2) For hosts in any given column of the periodic table, $|\psi_s(0)|^2$ decreases from the $3d$- to the $5d$-host.

Some of the deviations are probably accounted for by structural differences; e.g. Fe metal has a body-centred cubic structure while Ru and Os have a hexagonal close-packed lattice. Surprisingly, these influences are largely subordinated to the more general periodic relationship.

A more specific investigation concerns the magnetic properties of the alloy $Ni_3Al$ which shows a change from being strongly paramagnetic to weakly ferromagnetic as the nickel content increases. Furthermore, the presence of a minute proportion of Fe atoms can induce a giant magnetic moment by polarization of the host matrix. This can be as high as 39 $\mu_B$ per iron atom. Doping with $^{57}Co$ is a useful technique in that the impurity can be kept low

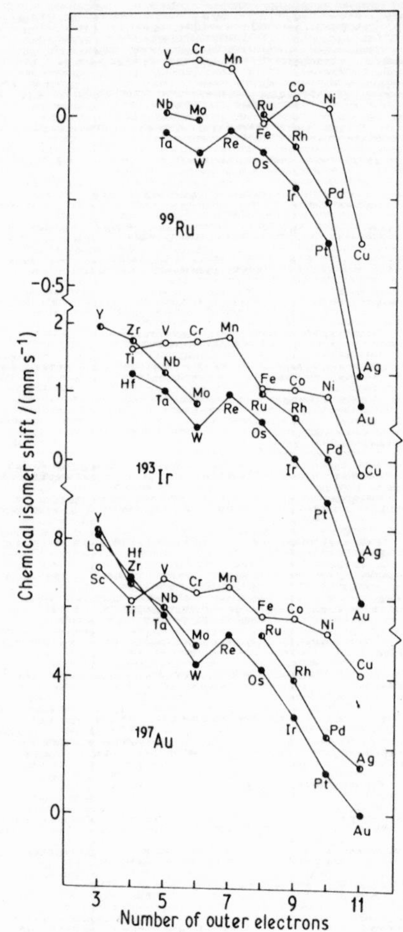


Fig. 10.2 The chemical isomer shifts of $^{99}$Ru, $^{193}$Ir and $^{197}$Au impurity atoms in the 3*d*-, 4*d*- and 5*d*-metals. The values were obtained from the emission spectra of $^{99}$Rh, $^{193}$Os and $^{197}$Pt, and are quoted with respect to Ru, Ir and Au metal respectively but in the sense of a conventional absorption experiment. $\delta R/R$ is positive in all cases, so that a more positive shift corresponds to a larger *s*-electron density at the nucleus. ([3], Fig. 1).

enough to obtain the maximum magnetic effect [4]. The $^{57}$Co emission spectra of phases with different compositions are shown in Fig. 10.3. They were recorded at room temperature where they are all paramagnetic.

The structure of $Ni_3Al$ is of the $Au_3Cu$ type with a face-

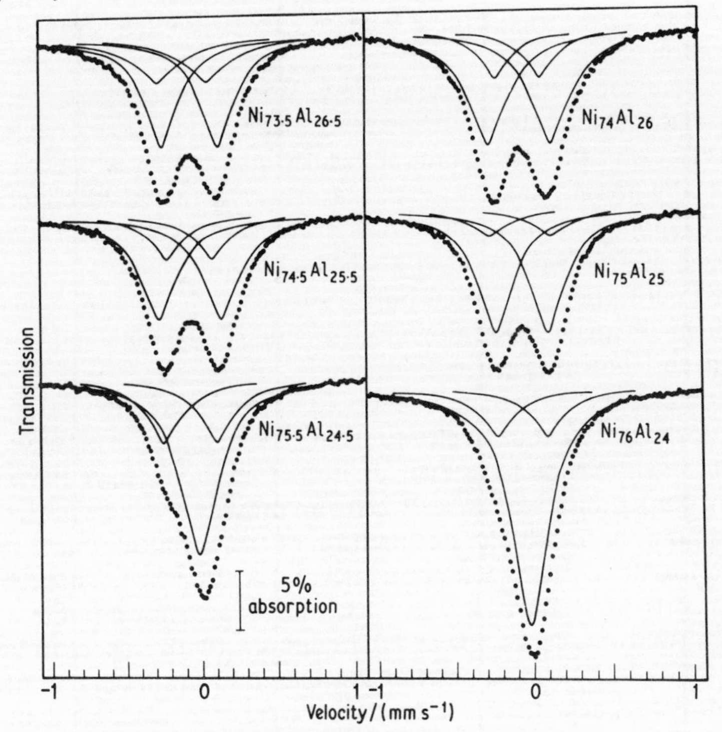


Fig. 10.3 The $^{57}$Co emission spectra at room temperature of doped alloys near the composition $Ni_3Al$. ([4], Fig. 1).

centred cubic lattice in which the Al atoms occupy the cube-corners and Ni atoms the face-centred positions. An Fe atom at the Al site should not show an electric field gradient because the site symmetry is cubic, but this is not true of the Ni site which has only axial symmetry. The dominant single-line resonance for a nickel content in excess of 75 at.%, and thus a deficiency of aluminium, may therefore be taken to indicate substitution at the Al sites. In the nickel deficient samples the converse takes place and iron is substituted at the Ni sites. The spectrum now shows a quadrupole splitting, but the shape is complex and appears to indicate the presence of a second weak doublet whose origin is not conclusively established. The magnetic spectra at low temperatures are particularly complex and will not be discussed here, but many of the apparent inconsistencies in earlier data for these alloys appear to stem from a tendency towards inhomogeneity in the alloy composition.

A comparatively new technique in Mössbauer spectroscopy is to isolate small molecules in an inert matrix and thereby obviate the effects of the strong intermolecular interactions which are normally present in the solid phase. This is achieved by condensing a vapour containing the atoms or molecules of interest (diluted at least 50-fold with an inert gas such as argon, xenon or nitrogen) under controlled conditions onto a surface at 4.2 K until a layer with an adequate absorption cross-section can be obtained. Isotopic enrichment is generally desirable.

Some of the early experiments of this kind were directed towards the isolation of single atoms of $^{57}Fe$ in matrices of argon, krypton and xenon [5]. Such an iron atom should have the free-atom configuration of $3d^6 4s^2 (^5D_4)$ and provide a valuable reference point in the absolute calibration of the chemical isomer shift scale for $^{57}Fe$. The spectra obtained at 4 K for $^{57}Fe$ in krypton with concentrations of Kr/Fe = 136 and 44 are shown in Fig. 10.4. The unsplit resonance line at $-0.75$ mm s$^{-1}$ (with respect to Fe metal at 300 K) corresponds to an isolated Fe atom on a Kr site in the close-packed cubic lattice. The weaker quadrupole doublet also recorded increases in intensity as the impurity concentration increases and can be attributed to an $Fe_2$ dimer. The splitting of 4.05 mm s$^{-1}$ is close to that expected for a single unpaired $3d$-electron. The $^5D$ atomic configuration of the free-atom will split in the axial crystal field produced by the adjacent Fe atom. The exact electronic ground state is not known, but the sixth electron (apart from the spherical $d^5$ half-filled shell) which produces the electric field gradient will occupy either $d_{xy}$ or $d_{x^2-y^2}$ or $d_{z^2}$. This could be resolved by determining the sign of $e^2qQ$.

The chemical isomer shift for the dimer is the same in both argon and krypton ($-0.14$ mm s$^{-1}$), and suggests that the dimer is a bound state with a constant internuclear separation. The value indicates an electronic configuration of about $3d^6 4s^{1.47}$; i.e. the large overlap of the $4s$-orbitals of the two atoms causes a large reduction in effective $4s$-density and a large increase in the shift observed.

Similar monomer/dimer behaviour has been recorded in $^{119}Sn$ [6]. Another tin species which has been observed is the isolated SnO molecule in matrices of argon and nitrogen [7]. The quadru-

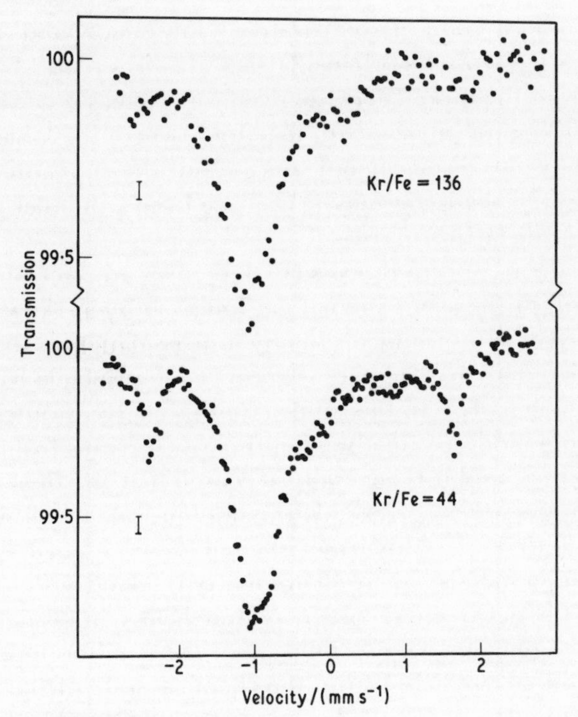


Fig. 10.4 The $^{57}Fe$ Mössbauer absorption spectrum of iron atoms isolated in a matrix of krypton at 4 K with relative concentrations of Kr/Fe = 136 and 44. ([5], Fig. 2).

pole splittings of 4.10 and 4.40 mm $s^{-1}$ respectively are unusually large for tin(II), and reflect the large imbalance in the occupation of the *p*-orbitals when only one covalent bond is formed. A more quantitative discussion of this splitting has not yet been devised.

## 10.2 Decay after-effects

In the majority of Mössbauer experiments the effects of the nuclear decay in the source can be ignored. This is because the source matrix has been carefully chosen to minimize any chemical after-effects. However, in insulating materials where the mobility of electrons is low it is not unusual to find that the decay of the Mössbauer precursor has a substantial effect on the chemical environment of the nucleus. The electron-capture, isomeric transition, and $\beta$-decays are of comparatively low energy and do not displace the atom from its original lattice site. They may be distinguished from the more energetic reactions such as $\alpha$-decay, Coulombic excitation and $(n, \gamma)$ *in situ* reactions, where the large recoil energy will cause displacement. As a result the emission of Mössbauer $\gamma$-rays in the latter may take place from a number of different sites in the lattice.

The electron-capture process (EC) removes an electron from an inner *s*-shell of the atom and is then followed by an Auger cascade. The net effect is to produce a highly ionized daughter state which for an isolated uncharged parent atom can in principle have a charge as high as +6. In the solid state this charge is quickly reduced to a more modest value by an influx of electrons from neighbouring atoms, and it is generally accepted that the final equilibrium charge-state is achieved within a time-scale much shorter than the lifetime of a Mössbauer excited level (i.e. $<10^{-8}$ s). For this reason, in metals, where the mobility of electrons is high, no effects of the decay are recorded. Nevertheless, in insulating solids, there remains the possibility that the equilibrium charge state is not that predicted from the chemical state of the parent atom, because the energetic Auger cascade may have had important effects on the immediate chemical environment. It is also possible, though more difficult to prove, that the atom is still in a metastable excited electronic state at the time of the Mössbauer measurement.

A recent experiment has clearly demonstrated the formation of multiple charge states by the EC decay of $^{57}$Co. $^{57}$Fe and $^{57}$Co atoms were isolated in solid xenon [8]. The $^{57}$Co emission and $^{57}$Fe absorption spectra of such a matrix containing ~1.1 at.% iron and ~10 $\mu$Ci $^{57}$Co are shown in Fig. 10.5. Only a minor proportion of the $^{57}$Co decays result in an isolated iron atom with the same $3d^64s^2$ configuration as the $^{57}$Fe. These give a resonance line at –0.76 mm s$^{-1}$ as described earlier (positive velocity defined in the sense of an absorption experiment). The chemical isomer shift of the major component in the emission spectrum at +1.77 mm s$^{-1}$ can be attributed to Fe$^+$ atoms with a $3d^7$ configuration. The large value of the shift is greater than found for $3d^6$Fe$^{2+}$ in for example FeF$_2$. This $3d^7$ configuration must be metastable as the free-ion ground-state is normally considered to be $3d^64s^1$. The latter configuration would give a very different chemical isomer

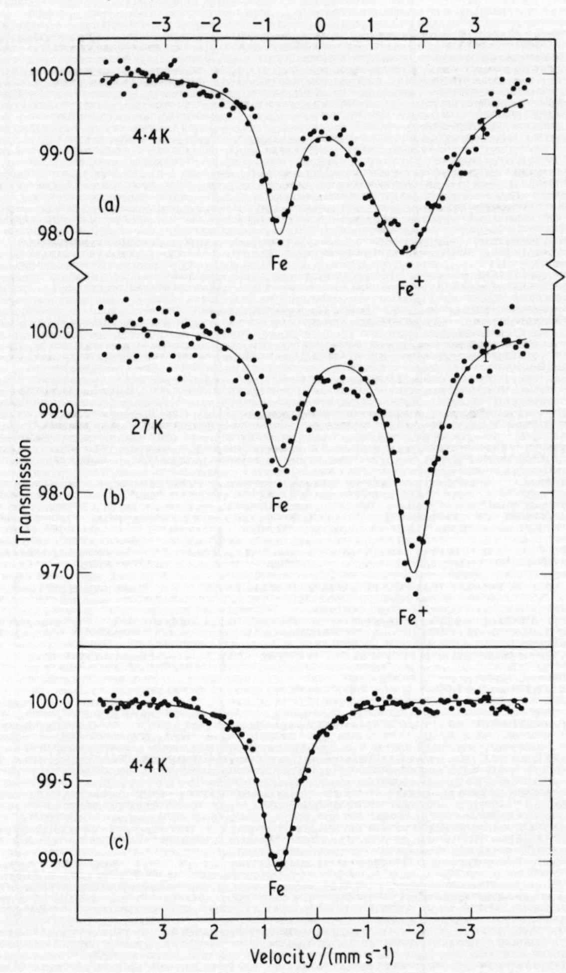


Fig. 10.5 The $^{57}$Co emission (top) and $^{57}$Fe absorption (bottom) spectra of a matrix of solid xenon containing ~1.1 at.% of iron and ~10 $\mu$Ci of $^{57}$Co. Only a portion of the electron-capture decays give the $3d^6 4s^2$ configuration of $^{57}$Fe which absorbs at $-0.76$ mm s$^{-1}$. The additional component at $+1.77$ mm s$^{-1}$ corresponds to $Fe^+$ ions in a metastable $3d^7$ state. The velocities in the source experiment have been reversed in sign to facilitate comparison. ([8], Fig. 1).

shift, and has been recently observed following ultraviolet irradiation of matrix-isolated $^{57}Fe$ atoms, and has a shift of +0.26 mm $s^{-1}$ [9].

The situation can be more complex when the decaying atom is in a chemically bound state. The EC decay of $^{57}Co$ in a simple ionic lattice compound involves insufficient energy to eject the atom from its lattice site. It might be anticipated that a daughter $^{57}Fe$ atom will be formed in the most stable oxidation state for the site occupied. This is probably what happens, but conflicting results have been found in many compounds (particularly oxides). This is believed to be because of the influences of non-stoichiometry and lattice defects on the charge-stabilization. It is not unusual to find that both $Fe^{2+}$ and $Fe^{3+}$ ions are formed in the same matrix, and $Fe^{+}$ or $Fe^{4+}$ ions are also sometimes observed. For example, the decay of $^{57}Co$ in CaO (NaCl face-centred cubic lattice) has been reported to result in $Fe^{+}$, $Fe^{2+}$ and $Fe^{3+}$ charge states [10]. The relative proportions in which different charge states are formed are often temperature dependent, and it is likely that the mobility of lattice defects and the ease with which electrons can be 'trapped' in the lattice are the primary factors.

Data for doped anhydrous halides appear to be more consistent than for the oxides. A typical spectrum from $^{57}Co$-doped $MgF_2$ at 295 K is shown in Fig. 10.6. The spectrum comprises two overlapping quadrupole doublets. That with a quadrupole splitting of 1.56 mm $s^{-1}$ and a chemical isomer shift of +0.49 mm $s^{-1}$ (in the absorption sense, relative to the $Na_4Fe(CN)_6.10H_2O$ absorber) corresponds to $Fe^{3+}$ (38% of the total area). The other doublet with a splitting of 2.82 mm $s^{-1}$ and a shift of +1.41 mm $s^{-1}$ corresponds to $Fe^{2+}$. It has been suggested in the past that $Fe^{3+}$ will tend to be formed if the cation size of the host is smaller than the size of the $Fe^{2+}$ ion (the ionic radii for $Fe^{2+}$ and $Fe^{3+}$ are about 76 pm and 64 pm respectively). However, no $Fe^{3+}$ is formed in $^{57}Co/MgCl_2$ where the $Mg^{2+}$ ionic radius is 65 pm. Recent results for series of related compounds such as $CoF_2$, $\beta$-$Co(OH)_2$, $CoCl_2$ $CoBr_2$ and $^{57}Co$-doped $MgF_2$, $ZnF_2$, $CoF_2$, $NiF_2$ suggest that the $Fe^{3+}$ charge state is preferentially stabilized in compounds with a large lattice energy [11]. This can be ascertained by reference to Table 10.1. In this context it is relevant to note the stabilization of $Fe^{4+}$ as well as $Fe^{3+}$ in $K_3\,{}^{57}CoF_6$ for similar reasons [12]. This compound has the cryolite structure which has a large value of

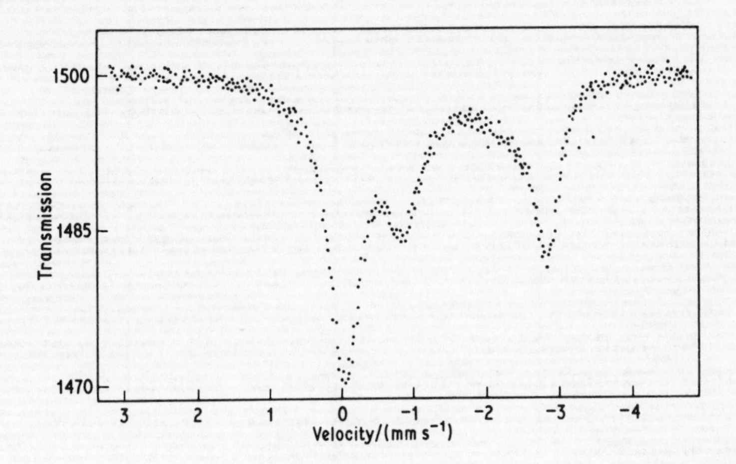


Fig. 10.6 The Mössbauer emission spectrum at 295 K for $^{57}Co$ doped into $MgF_2$. 38% of the decays produce an $Fe^{3+}$ ion, while the rest give $Fe^{2+}$. The spectrum has been reversed to resemble the equivalent absorption spectrum. ([11], Fig. 1).

Table 10.1 The fraction of $Fe^{3+}$ ions formed by electron-capture decay of $^{57}Co$ in simple ionic compounds

| Compound | Structure | Lattice energy /(kJ mol$^{-1}$) | % $Fe^{3+}$ |
|---|---|---|---|
| $NiF_2$ | rutile | 3080 | 59 |
| $MgF_2$ | rutile | 2930 | 38 |
| $CoF_2$ | rutile | 3060 | 32 |
| $ZnF_2$ | rutile | 2930 | $>0$ |
| $FeF_2$ | rutile | 2840 | 0 |
| $MnF_2$ | rutile | 2760 | 0 |
| $\beta$-$Co(OH)_2$ | $CdI_2$ | 2840 | 20 |
| $CoCl_2$ | $CdCl_2$ | 2610 | 0 |
| $CoBr_2$ | $CdI_2$ | 2590 | 0 |
| $MgCl_2$ | $CdCl_2$ | 2470 | 0 |

the Madelung constant and therefore a high lattice energy.

In co-ordination compounds with molecular ligands the effects of the $^{57}Co$ decay are more catastrophic. Hydrated $Co^{2+}$ salts show some $Fe^{3+}$ in the Mössbauer spectrum, but conversely some low-spin $Co^{3+}$ chelates generate a reduced $Fe^{2+}$ species. It is now widely accepted that the electrons and X-rays released in the Auger cascade often cause radiolysis of the surrounding ligands to produce ions and free radicals. The final charge-stabilization of the daughter iron atom is determined by the oxidizing or reducing properties of the radicals, and by the ease with which electrons can be 'trapped' nearby in the lattice. Thus the hydroxyl radical $OH^{\cdot}$ which will readily oxidize $Fe^{2+}$ to $Fe^{3+}$ is probably responsible for the $Fe^{3+}$ formed in hydrated compounds, while the organic radicals produced in $Co^{3+}$ chelates are highly reducing and will result in some $Fe^{2+}$ ions.

Convincing evidence in support of the autoradiolysis mechanism has been obtained by comparing the emission spectra from $^{57}Co$-doped $Co^{3+}$ chelates with the conventional absorption spectra of the isostructural iron compounds following intense irradiation with electrons [12]. Typical spectra for acetylacetone derivatives, $^{57}Co$:$Co(acac)_3$ at 77 K and $Fe(acac)_3$ irradiated at 80 K, are shown in Fig. 10.7. The similarity between the emission and absorption spectra in these and related complexes suggests that the species stabilized by the $^{57}Co$ decay are similar to those formed by external radiolysis. In the case of $^{57}Co$:$[Co^{III}(bipyridyl)_3](ClO_4)_3$, the pure compounds $[Fe^{II}(bipyridyl)_3](ClO_4)_2$ and $[Fe^{III}(bipyridyl)_3]$-$(ClO_4)_3$ are also known. Both the $^{57}Co$ decay and the external radiolysis of $[Fe^{III}(bipyridyl)_3](ClO_4)_3$ appear to generate the $[Fe^{II}(bipyridyl)_3]^{2+}$ species, but there may of course be significant differences in the detailed mechanism for formation of the reduced species in the two cases.

The radiation damage appears to be more severe in the more ionic complexes. In the instance of $^{57}Co$:$CoCl_2.2H_2O$ it has been demonstrated that the internal radiolysis produces a number of different defect species in both the $Fe^{2+}$ and $Fe^{3+}$ oxidation states [13]. This is particularly clear in the spectrum of the low-temperature antiferromagnetic phase where some of the defects do not participate in the magnetic superexchange and a 'paramagnetic' component is seen superimposed on the magnetic hyperfine splitting. The EC-decay is probably less destructive in the bipyri-

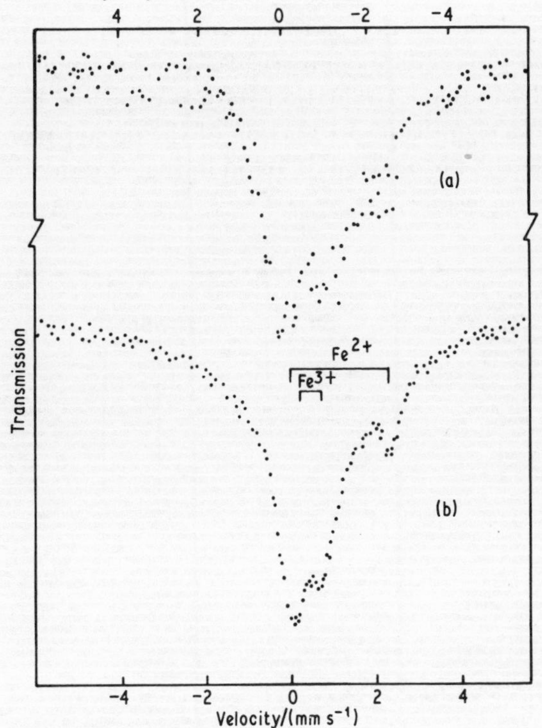


Fig. 10.7 The $^{57}$Co emission spectrum from (a) $^{57}$Co:Co(acac)$_3$ at 77 K and (b) the $^{57}$Fe absorption spectrum of Fe(acac)$_3$ at 80 K following irradiation with electrons at the same temperature. The sense of the velocity in the source experiment has been reversed. ([12], Fig. 4).

dyl derivatives because the ligand itself is more resistant to radiolysis, and it seems likely that none of the strong Fe-N bonds are ever ruptured.

The effects of a nuclear decay by isomeric transition (IT) are very similar to those of electron capture. A proportion of the decays result in an internal ionization (in this case by internal conversion of the $\gamma$-ray preceding the decay of the Mössbauer level) and a subsequent Auger cascade. The 23.8-keV level of $^{119}$Sn is populated by a highly converted isomeric transition from the 89.54-keV level of $^{119m}$Sn. A good example of after-effects is given by the decay of $^{119m}$Sn in tin(II) sulphate [14]. The $^{119m}$Sn:SnSO$_4$ emission spectrum and the $^{119}$Sn absorption spectrum of the same matrix are shown in Fig. 10.8. The sign of the velocity scale for the source experiment has been reversed to facilitate comparison. The tin atoms in the matrix are all initially in the +2 oxidation state as expected for tin(II) sulphate, but a significant proportion of the decays from the $^{119m}$Sn state result in the production of $^{119}$Sn in the +4 oxidation state.

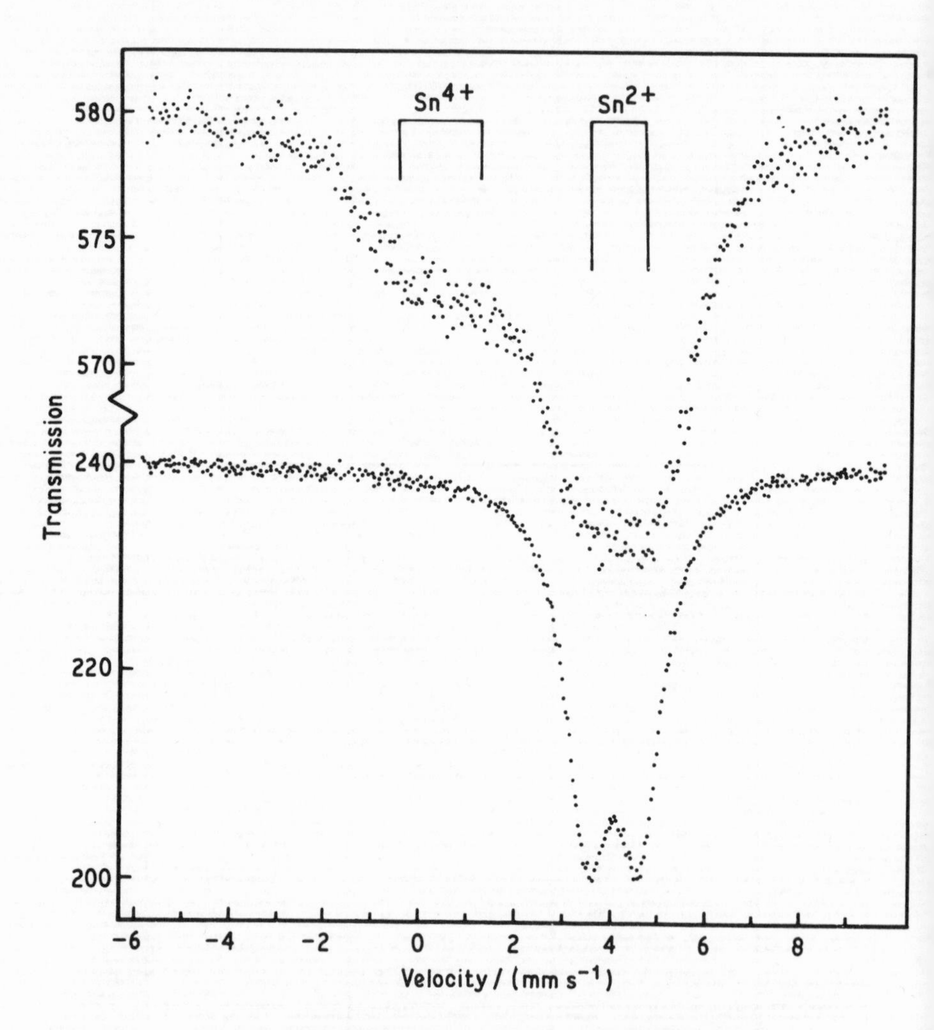


Fig. 10.8 The $^{119m}$Sn emission spectrum (upper) and $^{119}$Sn absorption spectrum (lower) from a matrix of $SnSO_4$ doped with $^{119m}$Sn. Note the production of some $Sn^{4+}$ species by the isomeric transition. The velocity scale of the source experiment has been reversed to facilitate comparison. ([14], Fig. 1).

The concentration of these defects is too low to register in the absorption spectrum. This oxidation is attributed to the consequences of an autoradiolysis process which accompanies the Auger cascade. However, decay after-effects are not very common in tin because the covalent bonds are usually stronger than in iron compounds.

If the primary decay is by $\beta$-emission, then the effects on the atomic environment are less drastic. The increase in the nuclear charge by $+e$ effectively increases the oxidation state of the atom. If the 'daughter' environment is isoelectronic and isostructural with a stable compound, then the co-ordinate geometry is unlikely to be affected. The primary influence appears to be the *chemical* stability of the daughter product in the environment of the parent lattice. In some instances it is possible to characterize previously unknown ionic species trapped in the host matrix. Thus the $\beta$-decay of $^{129}Te$ in $(NH_4)_2TeCl_6$ populates the 27.8-keV Mössbauer transition of $^{129}I$ [15]. A single-line $^{129}I$ emission is observed at $-6.08$ mm $s^{-1}$ relative to a $Zn^{129}Te$ source (i.e. at $+6.08$ mm $s^{-1}$ for an absorption). The regular octahedral $TeCl_6^{2-}$ anion appears to decay to an isostructural $ICl_6^-$ anion. There is also an alternative source precursor. $^{129m}Te$ decays by an isomeric transition to $^{129}Te$ prior to the ensuing $\beta$-decay. A source of $(NH_4)_2TeCl_6$ doped with $^{129m}Te$ gives the same single-line spectrum as from $^{129}Te$. In this instance neither the isomeric transition nor the $\beta$-decay cause molecular fragmentation, showing that the $ICl_6^-$ anion is comparatively stable.

Similar experiments have been used to prepare molecules of unknown xenon compounds. The 39.58-keV $I_g = \frac{1}{2} \rightarrow I_e = \frac{3}{2}$ resonance of $^{129}Xe$ can be easily observed using a source of $^{129}I$ in a matrix of NaI, $KIO_4$ or $Na_3H_2IO_6$, all of which show a single-line absorption with solid xenon or xenon hydroquinone clathrate as the absorber. In contrast, absorbers of $XeF_2$, $XeF_4$, and $XeO_3$ show well resolved quadrupole doublets with $\Delta = 39, 41$, and 11 mm $s^{-1}$ respectively because withdrawal of electron density from the xenon by the ligands causes an asymmetry in the valence orbital occupation.

Unexpectedly, the three sources mentioned above show small but significant chemical isomer shifts when compared with the same clathrate absorber, and it can be argued that the $\beta$-decay of the source generates a xenon species with some degree of chemical binding to the lattice. More dramatically, the decay of $^{129}IO_3^-$ in a matrix of $NaIO_3$ gives a spectrum with a clathrate absorber showing a quadrupole splitting identical to that from an NaI source and an $XeO_3$ absorber, so that it is an obvious inference that the $\beta$-decay has created an $\{XeO_3\}$ molecule trapped in the iodate lattice.

Similar experiments [16, 17] with $KICl_4.H_2O$, $KICl_2.H_2O$ and $KIBr_2$ sources show quadrupole split spectra attributable to trapped $\{XeCl_4\}$, $\{XeCl_2\}$ and $\{XeBr_2\}$ molecules, none of which are known as stable compounds but which may be presumed to be isoelectronic with $ICl_4^-$, $ICl_2^-$ and $IBr_2^-$ respectively. The observed splittings are $\Delta = 25.6$, 28.2 and 22.2 mm $s^{-1}$ respectively. Comparison of these values with those of $XeF_2$ and $XeF_4$ (see Fig. 10.9) show consistency with the predicted bonding in these molecules. $\{XeCl_4\}$ is square-planar, and $\{XeCl_2\}$ and $\{XeBr_2\}$ are linear molecules. In the tetrahalides, fluorine is more electron withdrawing than chlorine, and in the dihalides the order is fluorine > chlorine > bromine as predicted.

However, not all $^{129}I$ $\beta$-decay sources give a chemically bonded xenon atom. Both $^{129}I_2$ and $Cs^{129}IBr_2$ for example appear to break down to atomic xenon within the time scale of the Mössbauer event ($10^{-9}$ s). It should be emphasized that the $\{XeBr_2\}$ molecule in $KIBr_2$ may in fact have a lifetime of less than a microsecond.

An example of a change in oxidation state following $\beta$-decay is given by the decay of $^{193}Os$ in $K_2[OsCl_6]$ and $(NH_4)_2[OsCl_6]$ in which the $Os^{4+}$ has a low-spin $5d^4$ configuration. The decay populates the 73-keV level of $^{193}Ir$, and the oxidation state of the Ir daughter nucleus proves to be +4 ($5d^5$) [18]. This configuration is known to be stable in $K_2[IrCl_6]$ and $(NH_4)_2[IrCl_6]$, and presumably the $[IrCl_6]^-$ ion which is initially formed is rapidly reduced to the more stable $[IrCl_6]^{2-}$ configuration.

The effects on the Mössbauer spectrum of an energetic nuclear reaction such as $\alpha$-decay, Coulombic excitation by 25-MeV oxygen ions, or an $(n, \gamma)$ reaction are often surprisingly small despite the fact that the nucleus is displaced from its original site. In all cases the nucleus seems to reach its final lattice site in a time much less than the Mössbauer lifetime ($10^{-8}$ s), and in metallic matrices the spectrum recorded usually appears identical to that from a conventional absorption experiment. Few data are available for nonmetallic compounds, and it is in these that after-effects are likely to be severe. A reduction in the apparent recoilless fraction is

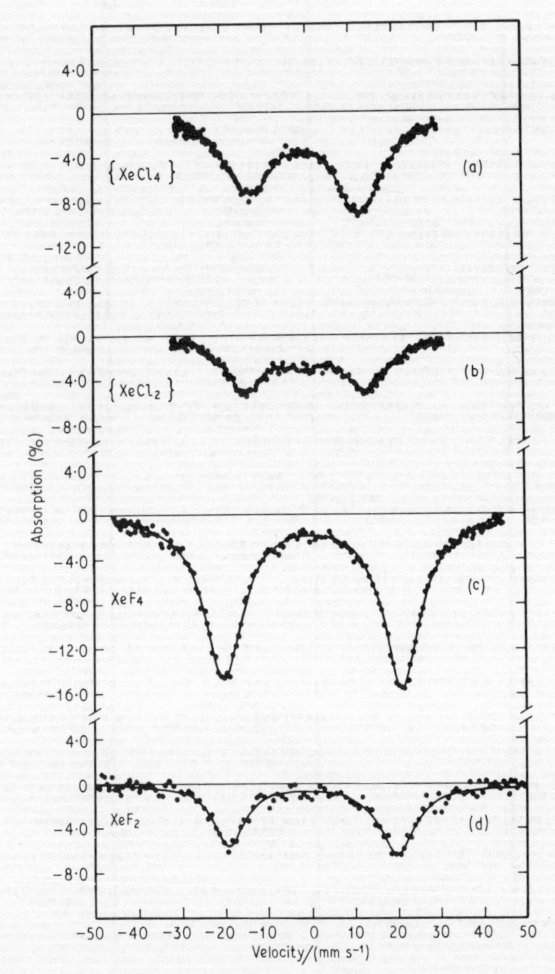


Fig. 10.9 The $^{129}$I Mössbauer emission spectra of (a) $\{XeCl_4\}$ trapped in $KICl_4$, (b) $\{XeCl_2\}$ trapped in $KICl_2$, and the $^{129}$Xe absorption spectra of (c) $XeF_4$ and (d) $XeF_2$. ([16], Fig.1).

sometimes found, although whether this is to be attributed to a residual kinetic heating or to lattice defects is not clear.

Nevertheless there is considerable scope for the study of radiation damage provided that the experiment is designed appropriately. A good example is given by data for thermal neutron capture by $^{192}$Os impurities in a matrix of $\alpha$-iron metal [19]. The $^{193}$Os daughter nucleus $\beta$-decays with a half-life of 31 *h* to the 73.1-keV state of $^{193}$Ir. Observation of the $^{193}$Ir Mössbauer emission spectrum thus records the environment of the $^{193}$Os atom long after any thermal disturbance has subsided. The irradiation was carried out at 4.6 K, and the Mössbauer spectrum recorded subsequently *without* any intervening rise in temperature showed an internal magnetic flux density of −138.6 T produced by magnetic alignment of the $^{193}$Ir within the iron lattice. The matrix was then annealed at successively higher temperatures by a heating/cooling cycle, and it was found that between 90 and 140 K the flux density (measured at 4.6 K) increased irreversibly to −140.8 T with a simultaneous narrowing of the resonance linewidth. Furthermore, these effects were independent of any change in the total neutron dosage, and were therefore related to the $^{192}$Os($n$, $\gamma$) capture event and *not* to general radiation damage of the iron lattice.

The recoil energy of the $^{193}$Os is estimated to be about 30 eV, which can be compared with estimates for the minimum energies required to produce displacement in iron along the (100), (110) and (111) directions of 17, 34 and 38 eV respectively. The observed reduction in the magnetic flux density in the unannealed state is believed to be due to a vacancy in the nearest neighbour iron site (following the general arguments of Chapter 8). The recoil of the ($n$, $\gamma$) reaction causes the $^{193}$Os atom to displace an iron-atom from a near-neighbour site, and thereby creates a vacancy at its original site. The displaced iron atom occupies a nearby interstitial position. This vacancy/interstitial defect can thermally anneal once the activation energy for migration of the atom of about 0.32 eV has been exceeded.

## References


[1] Steger, J. and Kostiner, E. (1973) *J. Chem. Phys.*, **58**, 3389.

[2] Abeledo, C. R., Frankel, R. B., Misetich, A. and Blum, N. A. (1971) *J. Appl. Phys.*, **42**, 1723.

[3] Wagner, F. E., Wortmann, G. and Kalvius, G. M. (1973) *Phys. Letters*, **42A**, 483.

[4] Liddell, P. R. and Street, R. (1973) *J. Phys. F*, **3**, 1648.

[5] McNab, T. K., Micklitz, H. and Barrett, P. H. (1971) *Phys. Rev. (B)*, **4**, 3787.

[6] Micklitz, H. and Barrett, P. H. (1972) *Phys. Rev. (B)*, **5**, 1704.

[7] Bos, A., Howe, A. T., Dale, B. W. and Becker, L. W. (1972) *Chem. Comm.*, 730.

[8] Micklitz, H. and Barrett, P. H. (1972) *Phys. Rev. Letters,* **28**, 1547.
[9] Data presented by Micklitz, H. and Litterst, F. J. at the International Conference on the Applications of the Mössbauer Effect at Bendor, September 1974.
[10] Regnard, J. R. (1973) *Solid State Comm.,* **12**, 207.
[11] Cruset, A. and Friedt, J. M. (1971) *Phys. Stat. Sol (B),* **47**, 655.
[12] Baggio-Saitovitch, E., Friedt, J. M. and Danon, J. (1972) *J. Chem. Phys.,* **56**, 1269.
[13] Friedt, J. M., Shenoy, G. K., Abstreiter, G. and Poinsot, R. (1973) *J. Chem. Phys.,* **59**, 3831.
[14] Llabador, Y. and Friedt, J. M. (1971) *Chem. Phys. Letters,* **8**, 592.
[15] Jones, C. H. W. and Warren, J. L. (1970) *J. Chem. Phys.,* **53**, 1740.
[16] Perlow, G. J. and Perlow, M. R. (1968) *J. Chem. Phys.,* **48**, 955.
[17] Perlow, G. J. and Yoshida, H. (1968) *J. Chem. Phys.,* **49**, 1474.
[18] Zahn, U., Potzel, W. and Wagner, F. E. (1973) Perspectives in Mössbauer Spectroscopy, ed. Cohen, S. G. and Pasternak, M. (1973) Plenum Press, New York, p. 55.
[19] Vogl, G., Schaeffer, A., Mansel, W., Prechtel, J. and Vogl, W. (1973) *Phys. Stat. Sol. (B),* **59**, 107.

CHAPTER ELEVEN

# *Biological Systems*

Some of the most interesting and stimulating applications of Mössbauer spectroscopy have been in the field of biological science. However, the practical and theoretical problems involved are probably more severe than in any of the topics covered in the preceeding chapters.

It is now recognized that many of the large protein molecules which control biological functions utilize the oxidation-reduction properties of a transition-metal atom, and that the complicated protein moiety functions largely in the role of ensuring steric specificity of the reaction. In very few instances has the detailed X-ray structure of a metallo-protein been determined. For this reason, any technique which can exclusively monitor the active metal-centre provides valuable information about the chemical role of a particular protein. For example, if the metal atom has a paramagnetic configuration, the electron spin resonance (E.S.R.) spectrum records the behaviour of the metal site to the exclusion of the protein bulk.

In a similar way, any protein containing iron can be studied using the $^{57}$Fe Mössbauer resonance which is also specific to the active site, but with the additional advantage that diamagnetic compounds, which are excluded from E.S.R. measurements, can also be studied. One of the major experimental difficulties lies in the relatively low iron content of proteins and the low natural abundance of $^{57}$Fe (2.17%). Fortunately the elements forming the amino-acid chains do not give a strong non-resonant scattering, but nevertheless with natural concentrations of iron the Mössbauer resonance is always comparatively weak. However, in many instances it is possible to cultivate organisms with a diet enriched in $^{57}Fe$, and the 40-fold improvement in the resonant cross-section which can be achieved results in an adequate absorption intensity. Alternatively, for some proteins it is possible to remove the iron and then to re-constitute the protein with enriched $^{57}Fe$ without affecting the biological activity.

The preparation of separated protein material is often a major task in itself, and particular care must be taken to ensure that accidental denaturing of the protein does not take place. Pure compounds in crystalline form are rarely obtainable, but experience has shown that satisfactory results may be obtained on frozen solutions of protein concentrates. Freeze-drying or lyophilization is also often used. Nevertheless, there remains the possibility of a conformational change in the solid so that the 'active' solution form is not observed.

The Mössbauer spectrum of the iron site reflects the ligand-field of the surrounding organic groups, and therefore the principles described in Chapters 4 and 5 are directly applicable. The major innovation is that individual iron atoms are now separated by distances of the order of 25 Å (2500 pm), and as shown in Chapter 6 this has an important influence on the relaxation properties of paramagnetic configurations. In such cases the Mössbauer spectra at low temperatures usually show a paramagnetic hyperfine splitting which slowly collapses with rising temperature. The spin-spin mechanism of electronic relaxation between neighbouring iron nuclei is comparatively ineffective, and in some instances is dominated by the very weak interactions between the electrons on the resonant atom and the nuclear spins of the ligand atoms. When this occurs, the zero-field Mössbauer spectra are extremely difficult to interpret. The application of an external magnetic field then results in a considerable simplification of the observed spectrum, from which many of the interaction parameters appropriate to the zero-field spectra may be obtained.

Some biologically active compounds such as vitamin $B_{12}$ (cyanocobalamin) contain cobalt, and although it is possible to dope with $^{57}Co$, this technique has not been widely exploited because of two major disadvantages. It is unlikely that the electronic configuration of the $^{57}Fe$ daughter nucleus will be related to that of the cobalt parent, and furthermore there is also a danger of local damage by autoradiolysis (see Chapter 10).

More detailed examples to illustrate these general principles are taken from the haemoproteins and ferredoxins.

## 11.1 Haemoproteins

The haemoproteins are well known as the oxygen carriers in blood. The oxygen is transported by chemical binding to an iron atom, which is itself in the centre of a planar porphyrin structure known as protoporphyrin IX (illustrated in Fig. 11.1). In the protein haemoglobin, the porphyrin is attached by co-ordination of the iron to a fifth nitrogen in a histidine unit of the protein chain below the plane of the four nitrogen ligands. The sixth co-ordination position of the iron on the opposite side of the porphyrin is thus left vacant, and forms a site for labile bond formation with small molecules.

It is possible to isolate compounds of iron with protoporphyrin IX, and although it might be expected that these would be model compounds for study of the haemoproteins themselves, this promise is not entirely fulfilled. The iron(III) chloride derivative of

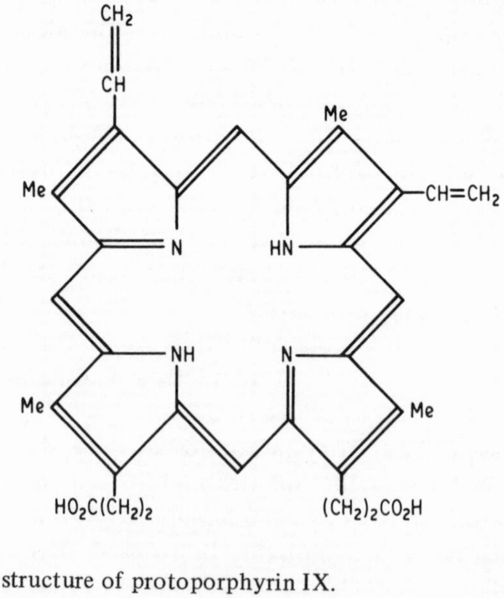

Fig. 11.1 The structure of protoporphyrin IX.

protoporphyrin IX is commonly referred to as haemin, and the reduced iron(II) derivative as haeme.

Data for haemin and other ring-substituted porphyrins with a selection of ligands in the fifth co-ordinate position show that the iron is normally present in the $S = \frac{5}{2}$ high-spin $Fe^{3+}$ state [1]. The unusually low chemical isomer shift and large quadrupole splitting (see Table 11.1) are not unexpected because of the substantial covalency in the bonds to nitrogen and the distorted symmetry of the square-based pyramidal geometry. The degree of distortion of the iron environment as monitored by an increase in the quadrupole splitting follows the sequence fluoro $<$ acetato $<$ azido $<$ chloro $<$ bromo.

These haemin derivatives are peculiar in that they show an unexpected temperature dependent effect. The spectrum at 4.2 K is a sharp quadrupole doublet, but as the temperature is raised the component at more positive velocity (which is the $m_z = \pm\frac{1}{2} \rightarrow m_z = \pm\frac{3}{2}$ transition because $e^2qQ$ is positive) broadens rapidly to give a very asymmetric spectrum. This effect is caused by a paramagnetic relaxation mechanism which is described in detail in Chapter 6 and illustrated there with data for $Fe(acac)_2Cl$ which has a similar co-ordinate geometry.

The bispyridyl haeme iron(II) derivative is an $S = 0$ low-spin compound, and the Mössbauer parameters are almost identical to those of the iron(III) derivatives because of the covalency effects.

Table 11.1 Mössbauer parameters for haemoprotein derivatives

| Compound* | $S$ | $T/K$ | $\Delta$ /(mm s$^{-1}$) | $\delta$(Fe) /(mm s$^{-1}$) |
|---|---|---|---|---|
| Haemin | $\frac{5}{2}$ | 4 | 0.83 | 0.36 |
| bispyridyl haeme | 0 | 4 | 1.14 | 0.36 |
| HbCO | 0 | 4 | 0.36 | 0.26 |
| $HbO_2$ | 0 | 1.2 | 2.24 | 0.24 |
| | | 77 | 2.19 | 0.26 |
| | | 195 | 1.89 | 0.20 |
| Hb (reduced) | 2 | 4 | 2.40 | 0.91 |
| HiF | $\frac{5}{2}$ | 1.2–195 | broadened | |
| HiCN | $\frac{1}{2}$ | 195 | 1.39 | 0.17 |

* The abbreviation Hb is used for an Fe(II) haemoglobin compound and Hi for an Fe(III) haemoglobin compound.

However, this diamagnetic configuration cannot show relaxation and therefore gives a sharp symmetrical spectrum at all temperatures.

The majority of the data on haemoglobin derivatives have been obtained using the blood from rats injected with ferric citrate solution containing iron enriched to 80% in $^{57}Fe$ [2]. It is not always easy to isolate pure compounds, and in this case spectra have usually been recorded in frozen solutions of red-cell concentrates or haemoglobin extracts. The oxidation state of the iron is very dependent on the nature of the sixth ligand. The carbon monoxide and oxygen derivatives of haemoglobin (HbCO and $HbO_2$) are both $S = 0$ low-spin iron(II) compounds. The spectra comprise a simple quadrupole doublet, and typical parameters are given in Table 11.1. The diamagnetic $S = 0$ state may be distinguished from paramagnetic configurations by the absence of any induced hyperfine splitting when a small external magnetic field is applied.

The applied-field method of determining the sign of $e^2qQ$ has been used to show that $e^2qQ$ is negative in $HbO_2$, but unfortunately the magnitude of $\eta$ is not known. The large quadrupole splitting and low chemical isomer shift of $HbO_2$ (compared for example to the bispyridyl haeme) show that the oxygen-iron bond is strongly covalent. A possible geometrical arrangement and bonding scheme [2] are shown in Fig. 11.2. The oxygen molecule is assumed to lie parallel to the porphyrin plane to give a strong overlap between the $d_{yz}$ metal orbital and the $\pi$*-antibonding orbital on the oxygen. The resultant separation of the molecular orbitals causes spin-pairing of the otherwise paramagnetic oxygen electrons to give the diamagnetic configuration observed. The initially non-bonding $d_{yz}$ electrons thus become intimately involved in the bonding. The molecular orbital $(d_{yz} + \pi_z^*)/\sqrt{2}$ has only half the character of a $d_{yz}$ orbital, and this results in an effective decrease in the occupation of this orbital. As a result, the principal axis of the electric field gradient tensor should lie along the $x$ axis of Fig. 11.2, and $e^2qQ$ should be large and negative (as observed experimentally). The quadrupole splitting is unusually temperature dependent for a diamagnetic configuration (see Table 11.1). This has been suggested to indicate a possible librational motion of the oxygen molecule at higher temperatures.

The paramagnetic derivatives show various configurations; e.g.

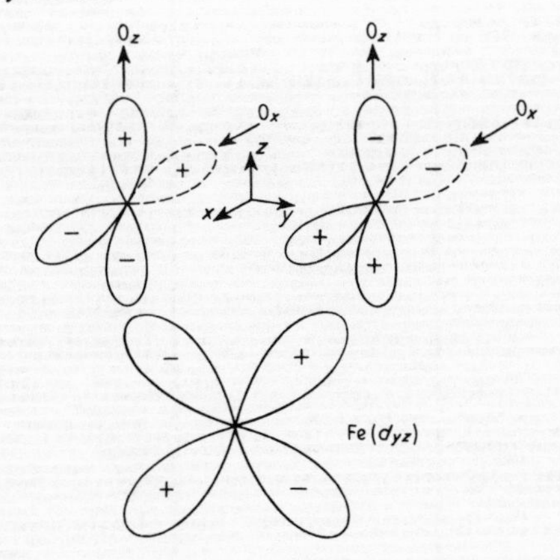


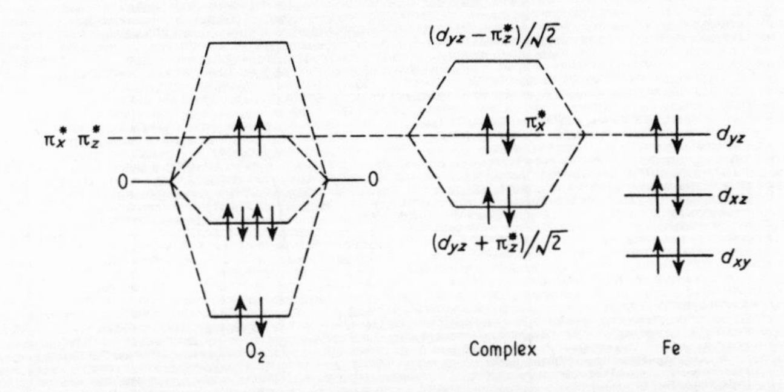


Fig. 11.2 A possible bonding scheme in $HbO_2$ showing how the interaction of the $\pi_z^*$ molecular orbital on the oxygen with the $d_{yz}$ orbital on the iron can remove the degeneracy and cause spin pairing in the bound state.

reduced haemoglobin is probably an $S = 2$ high-spin $Fe^{2+}$ compound, the fluoride (HiF) is an $S = \frac{5}{2}$ high-spin $Fe^{3+}$ compound, and the cyanide (HiCN) is an $S = \frac{1}{2}$ low-spin $Fe^{3+}$ compound. In all three the quadrupole-split spectra seen at high temperatures become grossly broadened on cooling by paramagnetic hyperfine splitting as the relaxation time increases. Equally complex behaviour results when an external magnetic field is applied. The resultant spectra are characteristic of each particular compound, and are a sensitive monitor of the iron environment. Successful solution of the appropriate theory can give highly detailed information. A good example is provided by the zero-field Mössbauer spectrum of haemoglobin fluoride at 1.2 K [3] which is shown in

Fig. 11.3. The $S = \frac{5}{2}$ $Fe^{3+}$ ion has the $S_z = \pm\frac{1}{2}$ Kramers' doublet lying lowest (see Chapter 6), and measurements in an applied magnetic field confirm this. However, in zero-field conditions the observed spectrum is clearly inconsistent with the prediction for an $S_z = \pm\frac{1}{2}$ electronic ground-state shown in curve (a) of Fig. 11.3. The discrepancy can be explained by invoking additional weak spin-interactions with the ligand nuclei which also induce electronic relaxation. Curve (b) was calculated including the effects of the fluorine nucleus, and is clearly a substantial improvement. A calculation including the fluorine and all four nitrogens is formidable, but curve (c) represents the effect of including only one additional $I = 1$ nitrogen nucleus in the calculation. The result is good evidence in support of these nearest-neighbour nuclear spin interactions. Future developments may lead to the determina-

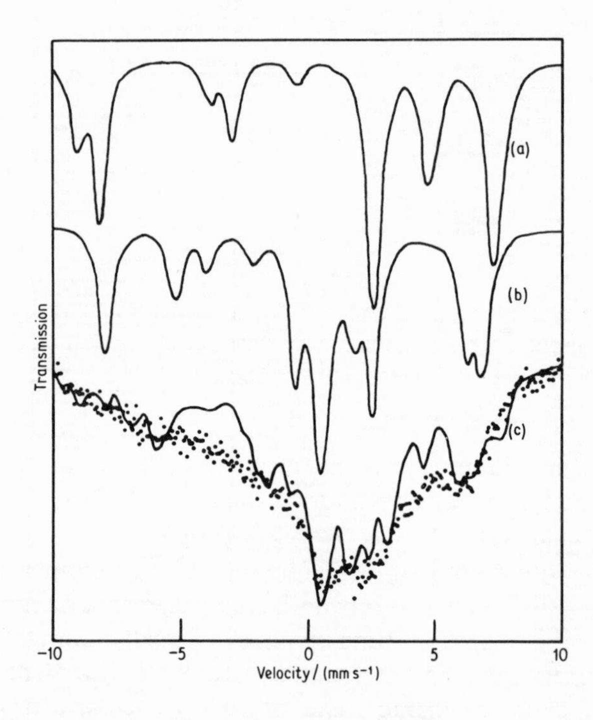


Fig. 11.3 The Mössbauer spectrum of haemoglobin fluoride at 1.2 K without an applied magnetic field. Curve (a) shows the predicted spectrum neglecting ligand-nucleus interactions. Curve (b) shows the predicted effect of including the transferred huperfine interaction with fluorine, and in curve (c) an additional $I = 1$ nucleus is introduced to represent nitrogen. ([3], Fig.1).

tion of the co-ordinate geometry in uncharacterized proteins.

The general principles described briefly here are also applicable to other haeme derivatives; e.g. the myoglobins involved in oxygen storage in muscle tissue, the cytochromes associated with the oxidation of nutrients in cells, and the peroxidase enzymes which activate hydrogen peroxide.

## 11.2 Ferredoxins

An equally important class of proteins occurring in bacteria and plants are the iron-sulphur proteins such as the rubredoxins and ferredoxins. They are associated with electron-transfer processes, and in particular with photosynthesis. Ferredoxins from the higher plants have a molecular weight in the region of 12 000–24 000 and contain two atoms each of Fe and labile sulphide (i.e. not attached to the amino-acid cysteine), and accept one electron upon reduction. The ferredoxins in photosynthetic bacteria contain eight atoms each of Fe and labile sulphide, accept two electrons per molecule, and are also more deeply involved in other chemical processes such as fixation of atmospheric nitrogen.

The Mössbauer spectra at 195 K of the oxidized and reduced forms of the plant ferredoxin from the green alga *Scenedesmus* are shown in Fig. 11.4 [4]. Similar spectra have been recorded in plant ferredoxins from the green alga *Euglena* and from spinach. The four resonance lines in the spectrum of the reduced form can be interpreted as two quadrupole doublets arising from $S = \frac{5}{2}$ high-spin $Fe^{3+}$ (inner doublet) and $S = 2$ high-spin $Fe^{2+}$ (outer doublet) in equal proportions. The parameters are given in Table 11.2. The chemical isomer shift of +0.56 mm $s^{-1}$ for $Fe^{2+}$ (with respect to Fe metal) is considerably lower than is usually observed because of the high covalency in a tetrahedral co-ordination to sulphur. The complicated spectra measured at low temperatures in an external magnetic field establish that the two iron atoms couple antiferromagnetically to give a net spin of $S = \frac{1}{2}$. The oxidized form of the ferredoxin contains two almost equivalent $S = \frac{5}{2}$ high-spin $Fe^{3+}$ cations which couple antiferromagnetically to give a diamagnetic ground-state. The iron atoms must therefore be very close together in the protein. Data from electron spin resonance measurements show that the co-ordination of the iron is tetrahedral. This active centre which is illustrated in Fig. 11.5

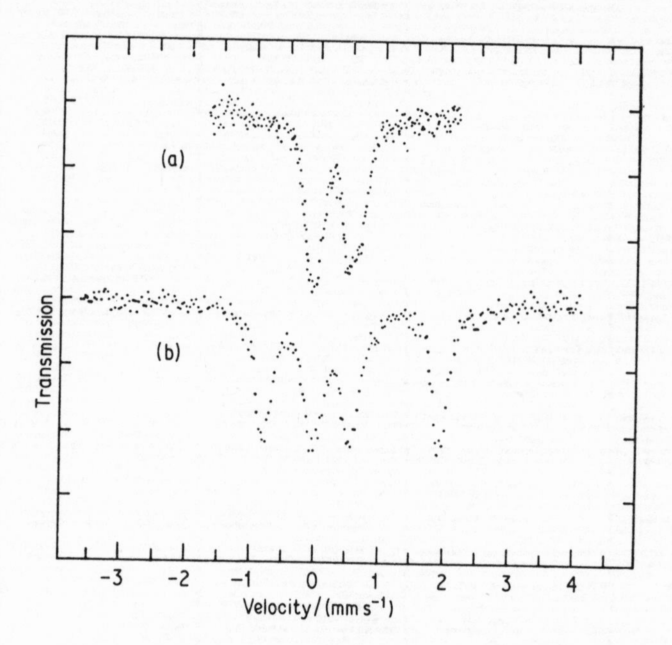


Fig. 11.4 Mössbauer spectra at 195 K of (a) the oxidized and (b) the reduced forms of *Scenedesmus* ferredoxin enriched with $^{57}Fe$. ([4], Fig. 6).

Table 11.2 Mössbauer parameters in ferredoxins

| Compound | $T/K$ | $S$ | $\Delta$ /(mm s$^{-1}$) | $\delta$(Fe) /(mm s$^{-1}$) |
|---|---|---|---|---|
| *Scenedesmus* ferredoxin | | | | |
| reduced form | 195 | $\left\{\begin{matrix}\frac{5}{2}\\2\end{matrix}\right.$ | 0.59<br>2.75 | +0.22<br>+0.56 |
| oxidized form | 195 | $\frac{5}{2}$ | 0.60 | +0.20 |
| *Chromatium* high-potential iron protein | | | | |
| reduced form | 77 | ? | 1.12 | +0.42 |
| oxidized form | 77 | ? | 0.79 | +0.32 |
| $(Et_4N)_2[Fe_4S_4(SCH_2Ph)_4]$ | | ? | 1.26 | +0.32 |

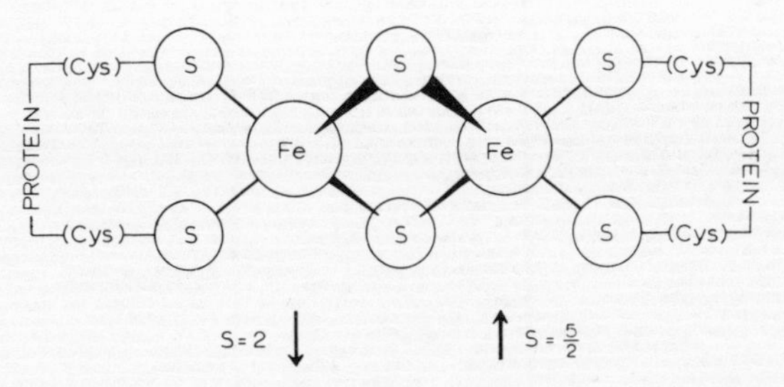


Fig. 11.5 Representation of the active site in the reduced form of a plant ferredoxin. The $S = \frac{5}{2}$ and $S = 2$ spin-states couple antiferromagnetically to give a resultant of $S = \frac{1}{2}$. In the oxidized form, both spin-states are $S = \frac{5}{2}$ and couple antiferromagnetically to give $S = 0$.

appears to be characteristic of all plant ferredoxins.

Much less is known about the bacterial ferredoxins. Recent X-ray studies have suggested the presence of a distorted cubic $Fe_4S_4$ cluster with iron and sulphur at alternate vertices. Such a cluster forms the active site of the high-potential iron protein from *Chromatium*, and two similar clusters are found in the clostridial ferredoxin from *Micrococcus aerogenes* (making a total of 8 Fe and 8 labile sulphide per molecule). Each iron is bonded to the polypeptide chain by a mercaptide sulphur on a cysteinyl residue.

The oxidized and reduced forms of *Chromatium* high-potential iron protein both give a simple quadrupole doublet at 77 K (see Table 11.2), and it appears that all four Fe atoms in the cluster are equally affected by the reduction [5]. Only the oxidized form shows paramagnetic relaxation effects at low temperature, although there is some evidence suggesting that there might be two structurally distinct pairs of iron atoms.

A similar $Fe_4S_4$ structural unit has recently been characterized [6] in the inorganic compound $(Et_4N)_2[Fe_4S_4(SCH_2Ph)_4]$. The Mössbauer spectrum between 4.2 K and 296 K is also a simple quadrupole doublet with similar parameters. Polarographic studies revealed a reversible one-electron transfer process, and there is good reason to believe that electron delocalization takes place within the $Fe_4S_4$ core. The compound may therefore be taken as a 'model' for the bacterial ferredoxins, the active centre of which can be represented as shown in Fig. 11.6.

The clear distinction between the Mössbauer spectra of ferre-

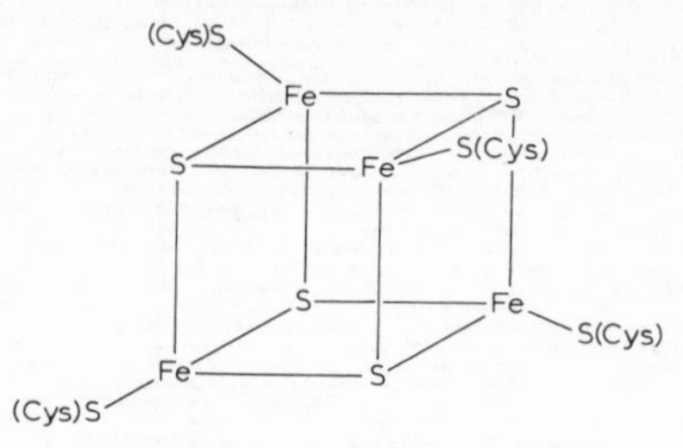


Fig. 11.6 Representation of the active site in the bacterial ferredoxins.

doxins from bacteria and from plants has proved useful in the biological classication of the intermediate algae. The blue-green algae and photosynthetic bacteria both lack a nucleus within the cell envelope and carry out photosynthesis throughout the cell. In contrast, the green algae and higher plants localize this function within a chloroplast inside the cell. However in other ways the biological activity of the ferredoxin in blue-green algae appears to be more closely related to the higher plants. The Mössbauer spectrum of a sample extracted from the planktonic blue-green alga *Microcystis flos-aquae* and chemically reconstituted with $^{57}$Fe enriched iron is identical with those described earlier from the higher plant forms *Scenedesmus* and spinach ferredoxins [7].

The establishment of this relationship of the blue-green algae ferredoxins to those of the higher plants provides useful evidence in support of the hypothesis that the chloroplasts in the cells of higher plants are the descendents of blue-green algae that lived symbiotically within the early plant cells.

## References

[1] Moss, T. H., Bearden, A. J. and Caughey, W. S. (1969) *J. Chem. Phys.*, **51**, 2624.

[2] Lang, G. and Marshall, W. (1966) *Proc. Phys. Soc.*, **87**, 3.

[3] Lang, G. (1968) *Phys. Letters,* **26A,** 223.

[4] Rao, K. K., Cammack, R., Hill, D. O. and Johnson, C. E. (1971) *Biochem. J.*, **122**, 257.

[5] Evans, M. C. W., Hall, D. O. and Johnson, C. E. (1970) *Biochem. J.*, **119**, 289.

[6] Herskovitz, T., Averill, B. A., Holm, R. H., Ibers, J. A., Phillips, W. D. and Weiher, J. F. (1972) *Proc. Nat. Acad. Sci. U.S.A.*, **69**, 2437.

[7] Rao, K. K., Smith, R. V., Cammack, R., Evans, M. C. W., Hall, D. O. and Johnson, C. E. (1972) *Biochem. J.*, **129**, 1159.

# *Bibliography*

The following selection of more general references are intended for the reader who wishes to pursue the subject further.

**Books**

*Mössbauer spectroscopy – an introduction for inorganic chemists and geochemists,* Bancroft, G. M. (1973) McGraw-Hill, London, 252 pp.

*Mössbauer effect and its applications,* Bhide, V. G. (1973) New Delhi, Tata McGraw-Hill.

*Mössbauer spectroscopy,* Greenwood, N. N. and Gibb, T. C. (1971) Chapman and Hall, London, 659 pp.

*Chemical applications of Mössbauer spectroscopy,* Goldanskii, V. I. and Herber, R. H. (eds.) (1968) Academic Press, New York, 701 pp.

*An introduction to Mössbauer spectroscopy,* May, L. (ed.) (1971) Plenum Press, New York, 202 pp.

*Der Mössbauer-Effekt und seine Anwendung in Physik und Chemie,* Wegener, H. (1965) Bibliographisches Institut, Mannheim, 214 pp.

*Mössbauer effect: principles and applications,* Wertheim, G. K. (1964) Academic Press, New York, 116 pp.

*Mössbauer effect data index 1958–1965,* Muir, A. H., Ando, K. J. and Coogan, H. M. (1966) Interscience, New York, (a catalogue of references up to early 1966).

*Mössbauer effect data index,* Stevens, J. G. and Stevens, V. E. (eds.) IFI/Plenum Data Corp., New York, (an annual publication listing references from 1969 onwards).

*Mössbauer effect methodology,* Vols. 1–8, Gruverman, I. J. (ed.) Plenum Press, New York, (an annual publication from 1965 containing a collection of contributed papers and reviews biased towards instrumentation)

*Perspectives in Mössbauer spectroscopy,* Cohen, S. G. and Pasternak, M. (eds.) (1973) Plenum Press, New York, 259 pp, (collected conference papers).

*Proceedings of the conference on the application of the Mössbauer effect – Tihany (Hungary), 1969*, Dézsi, I. (ed.) (1971) Akadémiai Kiado, Budapest, 803 pp.

*Mössbauer spectroscopy and its applications,* a Panel Proceedings of the International Atomic Energy Agency, Vienna (1972) 421 pp, (a collection of 16 review articles).

*Spectroscopic properties of inorganic and organometallic compounds*, vols. 1–7, Greenwood, N. N. (ed.) The Chemical Society, London, (containing an annual review of the literature from 1967 onwards).

**Reviews**


Bancroft, G. M. and Platt, R. H. (1972) Mössbauer spectra of inorganic compounds: bonding and structure, *Adv. Inorg. Chem. and Radiochem.*, **15**, 59–258.

Bearden, A. J. and Dunham, W. R. (1972) Iron electronic configurations in proteins: studies by Mössbauer spectroscopy, *Structure and Bonding*, **8**, 1–52.

Cohen, R. L. (1972) Mössbauer spectroscopy: recent developments, *Science*, **178**, 828–835.

Devoe, J. R. and Spijkerman, J. J. (1970) Mössbauer spectrometry, *Analyt. Chem.*, **42**, 366R–388R.

Friedt, J. M. and Danon, J. (1972) Applications of the Mössbauer effect in radiochemistry, *Radiochimica Acta*, **17**, 173–190.

Gibb, T. C. Applications of Mössbauer spectroscopy to organometallic chemistry. In *Spectroscopic methods in organometallic chemistry*, ed. George, W. O., p. 33–60, Butterworths, London.

Goldanskii, V. I., Khrapov, V. V. and Stukan, R. A. (1969) Application of the Mössbauer effect in the study of organometallic compounds, *Organometallic Chem. Rev. A*, **4**, 225–261.

Herber, R. H. (1967) Chemical applications of Mössbauer spectroscopy, *Progr. Inorg. Chem.*, **8**, 1–41.

Hobson, Jr., M. C. (1972) The Mössbauer effect in surface science, *Progr. Surface Membrane Sci.*, **5**, 1–61.

Johnson, C. E. (1971) Applications of the Mössbauer effect in Biophysics, *J. Appl. Phys.*, **42**, 1325–1331.

Kurkjian, C. J. (1970) Mössbauer spectroscopy in inorganic glasses, *J. Non-cryst. Solids*, **3**, 157–194.

Lang, G. (1970) Mössbauer spectroscopy of haem proteins, *Quart. Rev. Biophys.*, **3**, 1–60.

Maddock, A. G. (1972) Mössbauer spectroscopy in the study of the chemical effects of nuclear reactions in solids, MTP International Review of Science, Vol. 8, p. 213–250, Butterworths, London.

Parish, R. V. (1972) The interpretation of $^{119}$Sn-Mössbauer spectra, *Progr. Inorg. Chem.*, **15**, 101–200.

Reiff, W. M. (1973) Magnetically perturbed Mössbauer spectra of iron and tin co-ordination compounds, *Coord. Chem. Rev.*, **10**, 37–77.

Stevens, J. G., Travis, J. C. and DeVoe, J. R. (1972) *Mössbauer spectrometry*, *Analyt. Chem.*, **44**, 384R–406R.

Zuckerman, J. J. (1970) Applications of $^{119m}$Sn Mössbauer spectroscopy to the study of organotin compounds, *Adv. Organometallic Chem.*, **9**, 21–134.

# *Observed Mossbauer Resonances*

A total of at least 100 Mössbauer resonances have been detected in 83 isotopes of 44 elements. These are (energies in kev):-

$^{40}$K (29), $^{57}$Fe (14, 136), $^{61}$Ni (67), $^{67}$Zn (93), $^{73}$Ge (13, 68), $^{83}$Kr (9)

$^{99}$Tc (140), $^{99}$Ru (90), $^{101}$Ru (127), $^{117}$Sn (159), $^{119}$Sn (24), $^{121}$Sb (37), $^{125}$Te (35), $^{127}$I (58), $^{129}$I (28), $^{129}$Xe (40), $^{131}$Xe (80)

$^{133}$Cs (81), $^{133}$Ba (12)

$^{139}$La (166), $^{141}$Pr (145), $^{145}$Nd (67, 72), $^{147}$Pm (91), $^{147}$Sm (122), $^{149}$Sm (22), $^{151}$Sm (66), $^{152}$Sm (122), $^{153}$Sm (38), $^{154}$Sm (82), $^{151}$Eu (22), $^{153}$Eu (83, 97, 103), $^{154}$Gd (123), $^{155}$Gd (60, 86, 105), $^{156}$Gd (89), $^{157}$Gd (64), $^{158}$Gd (79), $^{160}$Gd (75), $^{159}$Tb (58), $^{160}$Dy (87), $^{161}$Dy (26, 44, 75), $^{162}$Dy (81), $^{164}$Dy (73), $^{165}$Ho (95), $^{164}$Er (92), $^{166}$Er (81), $^{167}$Er (79), $^{168}$Er (80), $^{170}$Er (79), $^{169}$Tm (8), $^{170}$Yb (84), $^{171}$Yb (67, 76), $^{172}$Yb (79), $^{174}$Yb (76), $^{176}$Yb (82), $^{175}$Lu (114)

$^{176}$Hf (88), $^{177}$Hf (113), $^{178}$Hf (93), $^{180}$Hf (93), $^{181}$Ta (6, 136), $^{180}$W (104), $^{182}$W (100), $^{183}$W (46, 99), $^{184}$W (111), $^{186}$W (122), $^{187}$Re (134), $^{186}$Os (137), $^{188}$Os (153), $^{189}$Os (36, 70, 95), $^{190}$Os (187), $^{191}$Ir (82, 129), $^{193}$Ir (73, 139), $^{195}$Pt (99, 129), $^{197}$Au (97), $^{201}$Hg (32)

$^{232}$Th (50), $^{231}$Pa (84), $^{234}$U (43), $^{236}$U (45), $^{238}$U (45), $^{237}$Np (60), $^{239}$Pu (57), $^{243}$Am (84)

# *Index*